Urban Sustainability

This book about urban sustainability encompasses the overarching core ideas of climate, cross-cutting concerns, advanced technologies, challenges, and solutions with reference to paradigms of climate change. It addresses green urbanism with detailed emphasis on urban infrastructure and explores strategies for managing common urban environmental issues. It unveils various social and cultural issues related to sustainable urban development, inclined towards equity, poverty, and public health, and offers guidance on how to create more sustainable, resilient, and liveable urban environments. It is aimed at graduate students and researchers in civil and environmental engineering.

Features:

- Explores a comprehensive account of urban-scale carbon neutrality.
- Provides thoughts about carbon neutrality's possible pathways, challenges, and related research.
- Covers the impact of carbon neutrality-based policies on people, their awareness, and their potential role in influencing policies.
- Includes biophilic cities, green gentrification, ghetto, and geospatial advancement in sustainability.
- Offers a combination of theory and case studies that highlight the versatility of the concepts discussed.

This book is aimed at graduate students and researchers in civil and environmental engineering.

Tanu Kumari completed her PhD at the Integrative Ecology Laboratory, Banaras Hindu University. Her doctoral research focused on riparian ecology under flood inundation scenarios. She has actively participated in national and international conferences, workshops, and seminars, and has published peer-reviewed research on sustainability, climate change, and ecosystem dynamics.

Pramit Verma is currently working in the field of urban ecology and sustainable energy. He has carried out postdoctoral work at Nicolaus Copernicus University and completed his PhD at Banaras Hindu University, India. He has worked in the field of urban climate disaster risk reduction and management and has published several research articles while participating in various national and international conferences, workshops, and seminars.

Pardeep Singh is currently working as an assistant professor in the Department of Environmental Science at the University of Delhi. He obtained his doctorate degree from the Indian Institute of Technology in 2017. The area of his doctoral research is the degradation of organic pollutants through various indigenous isolated microbes and by using various types of photocatalysts. He has published more than 35 papers in international journals in the field of waste management.

Urban Sustainability

Through the Perspective of Carbon Neutrality

Edited by
Tanu Kumari, Pramit Verma, and
Pardeep Singh

CRC Press
Taylor & Francis Group
Boca Raton London New York

CRC Press is an imprint of the
Taylor & Francis Group, an **informa** business

Designed cover image: Shutterstock

First edition published 2026
by CRC Press
2385 NW Executive Center Drive, Suite 320, Boca Raton FL 33431

and by CRC Press
4 Park Square, Milton Park, Abingdon, Oxon, OX14 4RN

CRC Press is an imprint of Taylor & Francis Group, LLC

ISBN: 978-1-032-83048-3 (hbk)
ISBN: 978-1-032-85416-8 (pbk)
ISBN: 978-1-003-51803-7 (ebk)

DOI: 10.1201/9781003518037

Typeset in Times
by codeMantra

Contents

SECTION III Tools and Techniques to Explore Urban Form-Energy Use-Carbon Emission Link

SECTION IV Cross-Cutting Issues

Preface

The threat posed by environmental degradation and climate change is universal and inescapable. Climate change is producing a multitude of environmental problems, many of which are concentrated in urban areas (heatwaves, urban floods, droughts, etc.), and these issues will likely get worse if cities keep expanding rapidly.

Environmental degradation and climate change pose a universal threat to humans as well as ecosystems. Climate change produces a multitude of challenges for urban areas such as heatwaves, urban floods, droughts, and human health issues. These will likely increase under the current course of urbanization. By 2050, two-thirds of the world's population is expected to live in urban areas. To bridge the gap between future complex environmental degradation and rapid urbanization, the notion of urban sustainability seems to be the most promising path. However, there are still several unresolved issues around this idea, which are difficult to integrate into urban planning and sustainable societies. Some of the nascent ideas that could coexist while developing sustainable cities, such as "green urbanism" and "ecosystem service evaluation", are understudied. Additionally, certain terms, for example, "green gentrification" and "biophilic urbanism" have entered the urban sustainability lexicon but have not yet been fully investigated or put into practice.

Cities can be considered the focal point for the efforts to slow down climate change effects because of their high levels of energy use and greenhouse gas (GHG) emissions, and direct and indirect impacts globally in the form of land use change, resource depletion, and waste production. Carbon neutrality, or net-zero emissions, is a state where carbon emissions produced are off-set as much as possible by their removal. This involves solutions drawing on technological, nature-based, community-based interventions, among several others. We have not, however, been as successful as we ought to be in achieving sustainable cities. Therefore, in order to attain carbon neutrality, we need to experiment with various ideologies like naturally based solutions or software-based tools (like CURB), green urbanism. Given this backdrop, this book offers various facets of urban sustainability, encompassing the overarching themes of core idea, climate, cross-cutting concerns, advanced technologies, challenges, and solutions while taking the paradigms of climate change into account. The discourse presented in this book ranges from exploring the resilience and adaptation pathways to deal with current climate-related challenges and prepare for future resilience.

Additionally, the chapters explore strategies for managing common urban environmental issues such as urban flood, urban heat island, and explore the potential of green urbanism, nature-based solutions and integration into smart-cities technology, green infrastructure and urban planning, water management, urban resilience, and adaptation strategies.

The concept of green urbanism has been explored in the first chapter, including its historical and conceptual evolution, its integration and discussion about smart city campaign in India. This is followed by chapters on transition from conventional urban paradigms to nature-based urbanism; and, nature-based solutions (NbS) to the impacts of extreme weather events, their role in increasing the social well-being, air-quality, and biodiversity. The need for integrating data-driven approach with NbS has been elaborated to increase its effectiveness in a smart-city paradigm.

This is followed by topics covering green infrastructure, techniques for sustainable urban planning and smart systems to mitigate the impact of climate-related disasters, and reducing the carbon footprint of cities. The theme of water management and urban resilience is further explored in the following chapter in the background of developing and implementing integrated water resource management, and adaptive governance. The theme of urban resilience is continued in the following chapter, where the role of cooperative governance, based on local action and regional-global partnerships, is explored. The role of adaptive capacity is described in a case study of agricultural sector in Nigeria, focusing on rural vulnerabilities and sustainable livelihoods. This book ends with a chapter on evaluating policies and strategies of climate-resilient public health policies for sustainable cities.

Beyond environmental concerns, this book also unveils various social and cultural issues related to sustainable urban development, inclined towards equity, poverty, and public health. Covering a wide-ranging view of the concept, principles, challenges, techniques, case studies, innovation, and attainable solutions in the context of urban sustainability under climate change, this book offers guidance on how to create more sustainable, resilient, and liveable urban environments for current and future generations with a carbon-neutral approach.

Contributors

Inobeme Abel is in the Department of Chemistry, Edo State University, Uzairue, Nigeria.

Bashir Adelodun is in the Department of Agricultural and Biosystems Engineering, University of Ilorin, Ilorin, Nigeria.

James Rotimi Adewumi is in the Department of Civil and Environmental Engineering, Federal University of Technology, Akure, Nigeria.

Fidelis Odedishemi Ajibade is in the Department of Civil and Environmental Engineering, Federal University of Technology, Akure, Nigeria.

Temitope Fausat Ajibade is in the Department of Civil and Environmental Engineering, Federal University of Technology, Akure, Nigeria.

Anil Barla is at the Institute of Environment and Sustainable Development, Banaras Hindu University, Varanasi, India.

Nazuk Bhasin is at the Institute of Environment and Sustainable Development, Banaras Hindu University, Varanasi, India.

Odozie John Chiwuzulum is in the Department of Agricultural and Development Economics, Ajayi Crowther University, Nigeria.

Işıl Demirtaş is in the Department of Economics, Giresun University, Turkey.

Abhulimen Fran Ehizokhale is at North Somerset Council, UK.

Ogbiti Temitope John is in the Department of Computer Science, Edo State University, Uzairue, Nigeria.

Chukwuemeka Kingsley John is in the Department of Construction and Civil Engineering, University College Jersey Highlands College, St Saviour Jersey, JE1 1HL, UK

Pankaj Kumar is in the School of Environmental Studies, Maa Shakumbhari University, Saharanpur 247120, India.

Sudhanshu Kumar is at the Institute of Environment and Sustainable Development, Banaras Hindu University, Varanasi, India.

Mathieu Nsenga Kumwimba is in the State Key Laboratory for Ecological Security of Regions and Cities, Research Center for Eco-Environmental Sciences, Chinese Academy of Sciences, Beijing 100085, China.

Kayode Hassan Lasisi is in the Department of Chemical Engineering, Faculty of Engineering and Design, University of Bath, Bath BA2 7AY, UK.

Khushi Mohan is in the Environment Management Division of Forest Research Institute, Dehradun, India.

Agbugui Marian Onwude is in the Department of Biological Sciences, Edo State University, Uzairue, Nigeria.

Adeniyi Osaimianmionmwan Arasomwan, is in the Department of Crop and Animal Science, Ajayi Crowther University, Oyo, Nigeria.

Okhamafe Elizabeth Princess is in the Department of Biological Sciences, Edo State University, Uzairue, Nigeria.

Gopal Shankar Singh is at the Institute of Environment and Sustainable Development, Banaras Hindu University, Varanasi, India.

Amit Kumar Tiwari is at the Institute of Environment and Sustainable Development, Banaras Hindu University, Varanasi, India.

Adamu Yunusa Ugya is in the Department of Environmental Management, Kaduna State University, Kaduna State, Nigeria.

Seda Yildirim is in the Department of Business Administration, Tekirdağ Namık Kemal University, Turkey.

Abdulhamid Yusuf is in the Department of Plant Science and Biotechnology, Federal University, Dutsin-ma, Katsina State, Nigeria.

Section I

Urban Sustainability and Climate Change

A General Overview

1 Green Urbanism
Principles and Purpose

Khushi Mohan

1.1 INTRODUCTION

1.1.1 What Is Green Urbanism?

Green urbanism (GU) refers to integration of urban planning with sustainable practises in all the possible aspects that are characteristic to an urban city such as energy, water and waste management, transport, green spaces and urban biodiversity, infrastructure, community health, living standards as well as governance, with an aim to create a culture of environmental compliance, equitable sharing of available resources as well as preserving them for future generations.

In other words, it is an interdisciplinary approach to build sustainable or environment-friendly community and lifestyle in order to reduce utilization of world's resources (Beatley, 2000). Lehmann (2011) states that GU is a coalition of efforts in reducing the use of water, energy and materials during the life cycle of a city or district since building a city involves considerable utilization of energy in obtaining the materials, transporting and assembling them to create infrastructures and their eventual recycling not forgetting the use requirement of water and energy in electricity, operation as well as maintenance needs of the infrastructure or city. All of the components that form an urban city need to be worked on collectively rather than separately for the city to be made become sustainable and efficient.

1.1.2 How Is It Important?

To understand the importance of GU, we first need to be aware about the global population scenario and the problems that are being created due to rapidly rising population as well as urbanization.

As projected by United Nations (2022) and later reported by Mcfarlane et al. (2023), total world population crossed 8 billion in November 2022, with India surpassing China and becoming the country with the world's largest population of 1.45 billion. United Nations has further projected the world population to increase to 8.5 billion, 9.7 billion and 10.4 billion in the years 2030, 2050 and 2100 respectively. By 2050, 68% of global population will be living in urban areas with a major rural to urban shift to be observed in Asia and Africa, with India adding upto 416 million urban residents (United Nations, 2018). Artmann et al. (2017) report that by the end of the 21st century, urban population percentage will reach 90% leading to

DOI: 10.1201/9781003518037-2

generation of new problems associated with booming urbanization along with the ones that are already present, such as urban sprawl. Such growing preference towards urbanization is creating pressure on present urban cities and it is further projected that globally, 1.2 million km^2 of new urban area will be built by 2030. Looking at the above-stated figures, we are headed towards a dangerous precedent as cities consume two-thirds of the global energy reserve and carry a share of more than 70% in greenhouse gas emissions (Mangweta et al., 2022).

Urban sprawl is the most prominent effect of rising population in a city as it occurs when such a city is no longer capable of sustaining its growing residents with essential services and amenities such as housing, schools, transportation, healthcare, sewage, clean air and water. It is an unplanned and unconstrained process of urbanization or urban expansion resulting from high-population growth as well as urban migration, targeted at the outer areas of an existing urban city. It demands greater land use, hence putting pressure on existing land resources and leads to unethical conversion of valuable land, whether agricultural or containing vegetation, resulting in loss of natural resources. Such resource constraints further cause socio-economic inequalities, as the high-income communities are able to secure the resources for themselves resulting in an unfair distribution of the crucial amenities within the urban poor. This in return perpetuates the poverty cycle and social disparity (Artmann et al., 2017; Khamdamov & Usmanov, 2024; Mangweta et al., 2022).

Urban sprawl also exacerbates other issues, including food insecurity, weak infrastructure, increased pollution (air, water and soil), inefficient energy use, increased automobile use, higher consumption and waste generation, reduced standard of living, health ill-effects and most importantly degrading environment. It further forms urban heat islands and contributes to climate change impact (Patacchini & Zenou, 2009).

The above challenges thus have increased the necessity for a greener approach in urban planning and development. The highly negative impact created collectively by rising population and uncontrolled urbanization can be reversed or mitigated by GU in the following ways:

- **Resource Conservation**: GU ensures proper management and conservation of resources such as energy, water and materials (Koller et al., 2022) by focusing on clean energy resources, efficient water and waste management as well as prioritizing the Rs: reduce, reuse, recycle, thereby lowering urban ecological footprint.
- **Mitigation of Climate Change**: The most important role of GU is mitigating climate change impact through green spaces that act as carbon sinks and regulate temperatures, and sustainable infrastructure including public transportation which aids in reducing CO_2 emissions. Saudi Arabia introduced the above strategies to implement its climate change mitigation plan (Abubakar & Dano, 2020; Gómez-Villarino et al., 2020; Sijakovic & Peric, 2021).
- **Public Health and Well-being**: Urban cities with green spaces, mixed-use of land and effective public transportation witness greater levels of

health and well-being. GU involves building a network of green spaces such as parks, for heat mitigation and reducing psychological stress and promotes recreation, physical activities, eco-friendly alternatives of movement such as cycling that further lessens air pollution (Stagno, 2021; Bosch & Sang, 2017).

- **Replenishing Biodiversity**: A sustainable urban design advocates for the creation and management of green infrastructure which is needed for conservation and enhancement of biodiversity and associated ecosystem services in the area. Sustainable urban development is closely associated with biodiversity conservation and public well-being (Farinha-Marques et al., 2011; Kowarik et al., 2020).
- **Economic Resilience**: The concept of GU exists to provide long-term solutions, one of which is building economic stability and resilience by shifting to circular economy which aims to reduce resource-use, waste generation and eventually environmental degradation. E. Hysa et al. (2020) report a positive and strong correlation between circular economy and economic development. As stated by Gürçam (2024), sustainable (or green) urbanism can strengthen economic resilience by minimizing carbon footprints and, improving disaster resilience as well as quality of life.
- **Social Inclusion**: Sustainable urban development leads to social inclusion by incorporating citizen participation, equitable sharing of environmental benefits and resources, and equal representation in planning (Taneja et al., 2022).

Cities today can move in a healthier direction if GU is applied in an efficient way such that its purpose is fulfilled and its principles are clearly defined which are going to be discussed ahead in this chapter, along with a closer look on the evolution of this concept, the current and future scenario as well as measures taken under the umbrella of GU to tackle the challenges created by climate change.

1.2 PURPOSE OF GREEN URBANISM

The purpose of GU is not merely the application of environment-friendly solutions but also building a holistic future involving social as well as economic sustainability. As per Cucoranu and Popescul (2023), *economic sustainability* comprises of a balanced, stable and inclusive economic growth that benefits the society, including individuals, and does not hamper the ability of coming generations to fulfil their own needs; whereas *social sustainability* includes equitable sharing of utilities, recreation, public safety, enhanced quality of life while ensuring judicial use of available resources.

GU as a concept has emerged to tackle the issues related to urbanization such that the development of a city is in harmony with nature along with focusing on the well-being of society and a healthy economy. As per Beatley (2000), a green urban city, that is, a city embodying GU ought to strive towards the following objectives:

- *It lives within its ecological limit*: It works to minimize its ecological footprint by monitoring factors that can heavily impact the environment, such as transportation, food production, energy generation and usage. It acknowledges its ability to affect the environment, other cities as well as planet as a whole.
- *It is green and employs ways to work with and for nature*: Instead of being the reason for environmental degradation, it should work towards propagating and benefitting nature by creating green spaces and urban forests central to the city, purifying air and water, restoring and nourishing existing biodiversity.
- *It follows a circular mechanism, taking inspiration from nature*: Nature is self-sustaining where waste is converted into useful input or resource. This principle can be applied to various functions of a city such as facility to convert municipal organic waste into fertilizer for food produced in the city.
- *It is self-sufficient* and manages its resources wisely.
- *It encourages its citizens to live a healthy and sustainable lifestyle* by providing them with the facility to walk, prefer public transportation, grow their own food, consume less, etc. and enabling greater satisfaction rooted in quality of life rather than quantity.
- *Its citizens enjoy high-quality life* surrounded by aesthetic and uplifting spaces and neighbourhoods filled with greenery and providing economically and socially inclusive housing.

Additionally, Lehmann (2011) has proposed a "triple-zero framework" which is a set of three goals on the basis of which the strategies under GU can be planned and executed. They are:

- *Zero use of fossil fuel-derived energy*
- *Zero waste*, which means the architecture is completely recyclable after its life-cycle ends
- *Zero or low-carbon emissions*

These objectives help envision a true green city, solving problems created via urbanization. It enjoys advantages such as higher living standards, maintained urban ecology and biodiversity, healthy and nurturing environment, efficient use of available resources, self-reliance in energy and food production, minimal pollution, proper waste management, reduced consumption and so on. It is also important to take note of the current and future world scenario and adding new goals to adapt to the changing times. These priorities would include:

- **Disaster Risk Management (DRM)**: It mainly constitutes measures for mitigation and adaptation, building urban resilience and preparedness to withstand and recover from unprecedented or slowly rising disaster risks including extreme weather events and climate change. Bignami (2020) highlights the need for sustainable urban planning to incorporate DRM to reduce the impact of rising disasters on cities and communities.

- **Green Innovation and Leadership**: For instance, green innovation comprises of developing new technologies, better recyclable materials and processes that reduce resource consumption and generation of waste whereas green leadership is about encouraging stakeholders to adapt to and promote sustainability (Zhong et al., 2023).

The one aspect that acts as a backbone in implementing all the goals of GU to create a well-planned and holistic city is *efficient policy and governance.* A governance that incorporates environmental considerations and sustainability in policies, forming decisions and their execution is known as **Green Governance**. It is different from a conventional governance framework that gives greater emphasis to economic growth and development at the expense of environment. Effectual green governance involves collaboration of various stakeholders like national and state governments, industries, NGOs, international organizations, educational and research institutions so as to leverage the expertise of each in addressing important environmental challenges.

In other words, the purpose of GU is to collectively build environmental, social and economic sustainability along with evolving suitably with changing times, supported by adept green governance. GU must be considered as the foundation for the future urban communities.

1.3 HISTORICAL BACKGROUND AND EVOLUTION

Before we dive into the history behind GU, let us first discuss how the term itself came into being. Dr. Timothy Beatley, a prominent researcher and author in the field of sustainable urban planning, coined the term "Green Urbanism" by introducing this approach in his book *Green Urbanism: Learning from European Cities* (Beatley, 2000). This term has been known by various other names in the past with the underlying principle of sustainability being in common.

We can trace the evolution of GU in both global and Indian context influenced by cultural, historical and socio-economic factors.

1.3.1 Global Context

The era of industrial revolution that began in Great Britain during the late 1700s and early 1800s and soon became globally prominent, introduced innovations and promised greater comfort in living but it was not without its detrimental impacts on the environment and public health which led to the emergence of GU worldwide.

Overcrowding and pollution that came with the industrialization inspired various authors and thinkers to write about and promote an idea of a city in the countryside, that is, among greenery and solitude, with Marsh (1865), Muir (1911) and Thoreau (1854) being examples of such works. Another early writing that can be linked to GU is Ebenezer Howard's *Garden City of Tomorrow* (1902) that was first published as *To-morrow: A Peaceful Path to Real Reform* in 1898. It emphasized a vision of a city wherein communities, industry and agriculture are present in proportion, lined by green belts, blending the benefits of rural and urban settlements. It further bloomed into the *Garden City Movement* in the 20th century (Howard, 1902).

1.3.1.1 Garden City Movement

The Garden City Movement is a remarkable urban design initiative of the late 19th century (1898) that was set forth by the ideas and efforts of Ebenezer Howard in the United Kingdom (Figure 1.1). It was in response to increasing migration rates towards cities due to industrialization, resulting in capitalism and overexploitation of nature. Howard laid down the planning principles of a garden city involving adequate green spaces, linked transportation, farms, healthcare, education and self-sufficient communities. Its physical layout is characterized by a central city that acts as an economic, cultural and social hub for its residents, surrounded by satellite cities that are self-sustaining and provide housing, local services as well as adjoined green spaces. These cities are connected to the central city via transportation increasing accessibility of the residents to all the available facilities (Howard, 1902; Nabila, 2021).

Based on Howard's concept, real-world garden cities came into fruition over the years (Nabila, 2021).

- *Letchworth*, Hertfordshire, United Kingdom being the first ever garden city designed by Raymond Unwin and Barry Parker for the working class in 1904.
- *Welwyn*, United Kingdom, the second garden city of England was founded in 1920, but it could not become self-sustaining.
- *Radburn*, New Jersey, USA, was built in 1928 to serve the emerging modern lifestyle before the Great Depression halted its development.
- As a fairly modern example, the Garden City principles were introduced in the planning of *Canberra Suburbs* in Australia in 2003.

The examples above prove that Howard's Garden City concept is still relevant to modern GU as it has been applied in several contexts, times and cities. It offers

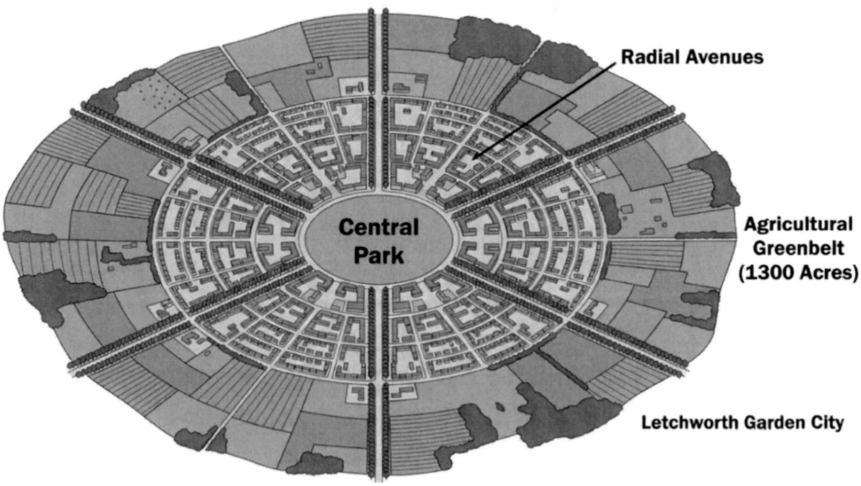

FIGURE 1.1 Garden City layout (reproduced from Howard 1902).

a framework for developing an urban space that brings the inhabitants and nature closer, is community-oriented and supports a stable economic growth, keeping sustainability at the forefront.

1.3.1.2 Important Periods of Green Urbanism in the West

Barton (2019) categorizes two crucial periods of urban greening: the first period is associated with development of green spaces during the 19th century for enhancing the quality of life and is considered to be the "forerunner of green urbanism in the late 20th century". The second period saw concrete efforts towards sustainable development in later part of the 20th century, that is, 1960s onwards, owing to the environmental consideration and movements, inspired by influential works such as Rachel Carson's *Silent Spring* (1962), Paul Ehrlich's *Population Bomb* (1968), *Design with Nature* by Ian McHarg (1969), Meadows et al.'s *Limits to Growth* (1972) and many more, addressing concerns like human-induced pollution, environmental degradation, population growth, resource limit and so on (Carson, 1962; Ehrlich, 1968; McHarg, 1969; Meadows et al., 1972).

The building awareness and pressures of the movement soon led to the first ever world conference that discussed the environment and related socio-economic issues, also known as the United Nations Conference on the Human Environment, which held in Stockholm, Sweden in 1972. The Stockholm Conference led to the establishment of the United Nations Environment Programme, policy reforms and a "new focus for urban planning" (Barton, 2019). Another major contribution in the environmental movement was the famous Brundtland Report (1987) also named as "Our Common Future", published by World Commission on Environment and Development, that introduced the term *Sustainable Development* and defined it as "development that meets the needs of the present without compromising the ability of future generations to meet their own needs" (Brundtland & Khalid, 1987). Other important meetings and decisions followed during that time and into the 21st century as well, with a more nuanced focus on the issues of the environment rather than being completely human-centred (as in the Stockholm Conference) including the Earth Summit (1992), Kyoto Protocol (1997), etc.

Barton (2019) fairly dissects that, as a part of the second period of urban greening awareness, works of planners and authors such as Timothy Beatley, Steffan Lehmann and Peter Newman greatly contributed in developing and furthering the concept of GU with Beatley establishing the term in 2000. These works focussed on integrating environment more closely in urban planning and architecture, forming principles required to bring the vision of a city embodying GU to reality. It involved the support of community across all aspects of living such as housing, transportation, water, waste, materials, biodiversity, energy, public space, climate change and so on (Beatley, 2012; Beatley & Newman, 2013). Lehmann (2011) efficiently summarized those principles that will be discussed ahead.

The concept besides being called as GU also began to be introduced as *sustainable development, sustainable cities* (Beatley, 2000), *green city* (Karlenzig et al., 2007), *sustainable urbanism* (Farr, 2008), *eco-cities* (Lehmann, 2011), *compact cities* (Jenks & Burgess, 2002), etc. involving the same underlying principle but with increasing understanding towards factors required to make a city sustainable in a holistic way. Lehmann and Mainguy (2010) defined the term *New Urbanism* as an "urban design movement" which promoted a compact city to prevent urban sprawl whereas Beatley (2000) described GU as a "different kind of new urbanism" which is much more ecologically inclined.

The *modern green urbanism* of the 21st century focuses on creating compact cities that are close to nature and also resilient to deal with emerging issues such as climate change and associated disaster risks, incorporating international decisions and agendas (Matan & Newman, 2015).

1.3.2 Indian Context

Ancient India has always been ahead of most other countries in terms of knowledge and concepts that have been re-discovered much later and popularized with another name, usually under the guise of so-called Western findings. Sustainable development has been a Vedic philosophy long before it was given heed by the world at the Stockholm Conference. It is surprising how long the world took to grasp what Indian population had already been practising in its day-to-day life, deriving the values and wisdom from the Vedas. Rai and Rai (2024) report that the Vedic texts advocate such progress which does not "compromise the needs of future generations".

1.3.2.1 Ancient and Medieval Examples of Sustainability

Architectural spectacles discovered or constructed during ancient and medieval India are important practical applications of GU during that time. Sustainable lifestyle practised then is still relevant for the current scenario.

1.3.2.1.1 Indus Valley Civilization

IVC had advanced water and waste management systems, especially in cities like Harappa and Mohenjodaro, that encompassed reservoirs and drainage systems respectively. Along with this, A water reservoir was discovered in Dholavira, indicating saving and storage of water. Sustainable agriculture was practised by employing crop rotation and maintaining soil fertility in organic ways. Large granaries found at IVC sites were used to store the surplus crop and manage agricultural resources (Figure 1.2). Artefacts depicting an animal or other element of nature indicated a deeper connection to environment (Bhardwaj, 2024; Kineman & Anand, 2015).

1.3.2.1.2 Agra Fort

Agra Fort is a well-known medieval architecture with its ingenious water management techniques (Figure 1.3) and systems such as *baoli* (stepwell), water reuse mechanism, fountains that irrigated the gardens using recycled wastewater, moat that acted as a storage pit for groundwater recharge and treatment of wastewater before disposal.

The above figure illustrates a well-managed hierarchical system of water use that worked with the natural flow of water. Clean water was drawn from the Yamuna

FIGURE 1.2 Southern (a) and Eastern (b) Reservoirs of Dholavira, Gujarat (Singh et al., 2020).

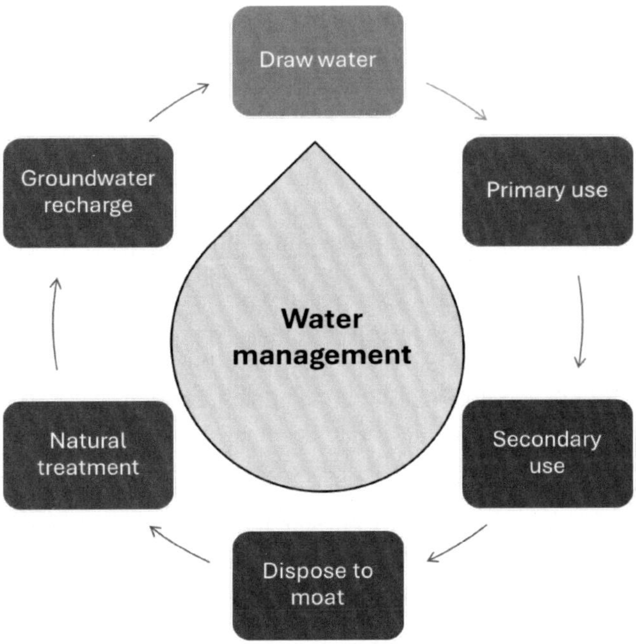

FIGURE 1.3 Water Management System at Agra Fort.

River or groundwater from the baoli using a Persian wheel operated by horses or bulls. This was the only stage requiring mechanical power. Water then flowed down for primary (drinking, cooking) and secondary use (cleaning, irrigation) and then would be disposed to the moat where it was naturally treated and used for groundwater recharge. The water flow, except in the first stage, was maintained by gravity thus involving no energy use in its distribution.

The fort's architecture also enabled natural light and ventilation promoting efficient use of energy, integration of green spaces that benefitted biodiversity and living standards, and *nalis* then ran in every room containing moving water for temperature regulation. The fort was self-sustaining with communities and local markets nearby reducing the need of transportation (Zuberi, 2017). The knowledge and wisdom of the period was way ahead of its time with more such examples present that can truly inspire architectural advancements today.

1.3.2.2 Political and Literary Examples

Indira Gandhi, the then Prime Minister of India, established a link between poverty and environmental degradation stating, "Aren't poverty and need the biggest polluters?" in her speech at the Stockholm Conference 1972, which has been open to various interpretations by environmentalists. One such crucial conclusion is that poverty can be considered both – a cause and consequence of environmental deterioration: cause being the lack of awareness and resorting to unsustainable practices due to necessity, and consequence being the further exacerbation of poverty due to decreasing natural resources. The speech also emphasized inclusion of poverty alleviation in sustainable development strategy (Mudaliar & Kashwan, 2024).

The idea of simple holistic living in India was advocated by Mahatma Gandhi, focusing on living below means and in harmony with nature to conserve resources and an economic growth that is localized, equitable and sustainable (Sahoo, 2020). He believed that such an approach would make a nation's economy steady and permanent rather than exponential and hence, short-spanned. This concept inspired the work of J. C. Kumarappa, an Indian economist, titled as *Economy of Permanence* describing an economic model based on principles such as ecological balance, non-violence and self-sufficiency (Kumarappa, 1945).

E. F. Schumacher, a British economist, influenced by both of the above personalities produced the crucial work *Small Is Beautiful: A Study of Economics As If People Mattered* stressing that reckless exploitation of natural resources in lieu of materialism is not the solution to economic development and the focus should be shifted on a growth based on environmentalism (Schumacher, 1973). His work came to be highly regarded in the West.

Advocacy of sustainability by important figures such as mentioned above is essential to spread the awareness among citizens as well as prominent stakeholders. It holds the power to alter policies and decision-making towards a positive direction.

1.3.2.3 Initiatives in Modern India

In India, the colonialism and post-independence era witnessed heavy industrialization and exploitation of natural resources. Based on the projections by the United Nations and the Ministry of Statistics and Programme Implementation, India's present population is 1.46 billion approximately making it the world's most populous country. The country's urban population in 2011 was recorded to be around 377 million and is further projected to reach around 600.9 million by 2036, marking a 59.4% increase (Handbook of Urban Statistics, 2022). This rapid urbanization is because of rural-urban migration as well as natural population growth, fuelled by employment opportunities and better living standards. But this has led to other challenges such as slum expansion, pollution rise, traffic congestion and urban sprawl.

To tackle such issues seen in Indian urban cities and introduce sustainability, various urban development projects and programs have been initiated by the Government of India as well as other organizations, some of which are mentioned below:

- **AMRUT Yojana (Atal Mission for Rejuvenation and Urban Transformation):** Earlier known as The Jawaharlal Nehru National Urban Renewal Mission (JNNURM), this program was launched in 2015 by GOI to improve urban infrastructure, governance, community participation and basic amenities for the poor and disadvantaged, in specific groups of large cities throughout the country. Its main objectives involve ensuring water supply and sewage disposal in every household, integration or maintenance of public green spaces, encourage public transport and develop proper drainage systems to prevent flooding (Smith & Pathak, 2018).
- **LEED-India (Leadership in Energy and Environmental Design):** LEED is a globally recognized green building certification and rating system that is also preferred by India's private sector. It comprises of standards for sustainable design and construction methods of buildings and awards points to projects based on sustainable development of site, materials selection, water conservation, energy efficiency and quality of indoor environment. ITC Hotel chain in India was awarded the highest LEED rating, that is, LEED Platinum in 2011, making it a "leadership in energy and environmental design" (Smith & Pathak, 2018).
- **GRIHA (Green Rating for Integrated Habitat Assessment):** GRIHA is a quasi-public green rating system established by The Energy and Resources Institute (TERI) and the Ministry of New and Renewable Energy. It is an indigenous alternative to the Western rating systems to cater to the climatic and geographical context of India. It assesses factors like water usage, waste generation, energy consumption, materials, incorporation of renewable energy, etc. during the entire life-cycle of a building. In 2010, it was made mandatory for all new government buildings to have a minimum of 3-star GRIHA rating, with 1–5 being the rating range (GRIHA, 2021).

There has been a shift towards more environmentally oriented practices in the country, with GU spreading globally. Major urban cities in India, such as New Delhi, Chennai and Pune have been incorporating greening practices to tackle issues such as air pollution, urban heat island effect and resource depletion (Imam & Banerjee, 2016). The low-carbon city concept is being applied for reducing carbon emissions and promoting renewable energies (Sahni & Aulakh, 2014) whereas the Smart City Mission launched in 2015 has been promoting modernization of existing cities by improving facilities, infrastructure, technologies and developing green economy (Bhattacharya, 2021). Rise in green buildings and spaces has been another implementation of GU in Indian cities (Ramesh & Nijagunappa, 2015) among other initiatives.

Along with the application of abundant traditional knowledge in India's culture and history, the country is also adjusting to the global standards and policies with time, making progress towards incorporating the principles of GU in its growing urban settlements while there is always a room for improvement.

1.4 CORE PRINCIPLES OF GREEN URBANISM

To achieve the vision of a city, as created by various thinkers and planners in the line of GU, we need a set of clear principles that will align our actions towards the desired result. For a better understanding, these principles can be considered as strategies, guidelines or a framework that breaks down the overwhelming concept of GU into executable components.

The principles described in Figure 1.4 are *universal* and *holistic*, each being essential in nature, based on the work by Lehmann (2011) with added viewpoints of Newman (2010) and other authors. Lehmann has given 15 principles based on the "triple-zero framework" of zero waste, zero use of fossil-fuel energy and zero emissions, that are required to be adapted as per the specific location, its characteristics and requirements such as the scale of development, economy, available technology, goals and so on.

The core principles of GU described below may be overlapping in a few areas but attempt to highlight ways to reach the goal of sustainability.

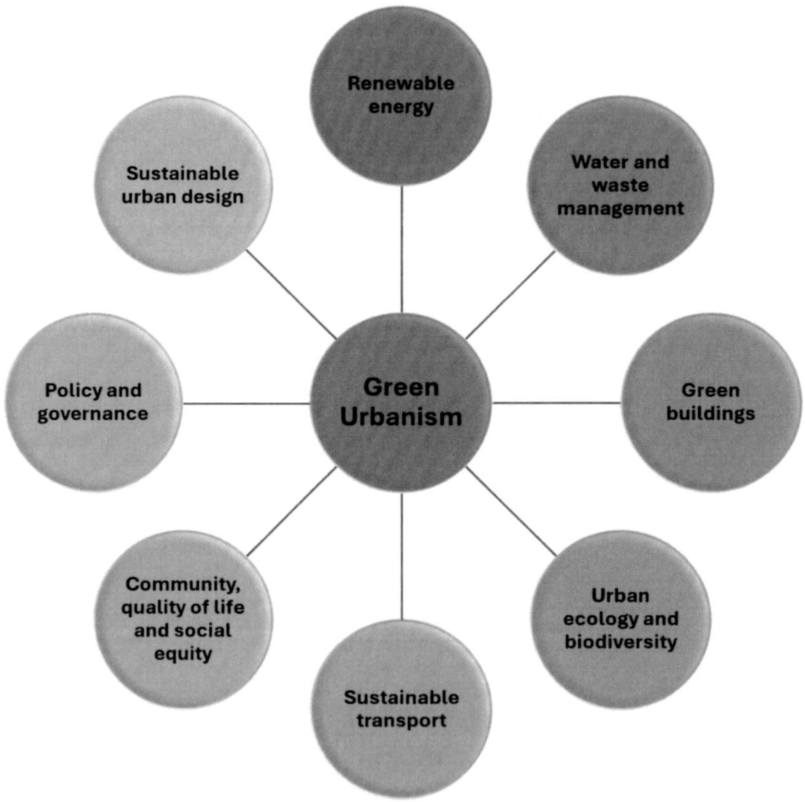

FIGURE 1.4 Core Principles of Green Urbanism.

1.4.1 Sustainable Urban Design

Sustainable urban design involves planning in every aspect of a city such as infrastructure, transportation, green spaces but it is mainly about adjusting the urban layout and developmental plans according to the characteristics of the specific city as well as changing times. Mentioned below are important components that help form a beneficial urban design.

1.4.1.1 Climate and Context

The climatic conditions and context of the site need to be considered in any project design, involving temperature, humidity, precipitation, topography, resources available, sunlight, pollution levels, and so on. The goal is to efficiently utilize what is available to attain maximum possible sustainability. The potential of the location needs to be recognized and taken advantage of, tailoring the urban design accordingly.

1.4.1.2 Compact City and Mixed Land Use

As discussed before, urban sprawl is the expansion of a city horizontally causing unsustainable practices such as higher consumption of land, excessive resource-use, destruction of agricultural and vegetation land, increase in number of automobiles and resulting enhanced pollution levels, as well as other economic and social issues.

A *compact city* is the opposite of urban sprawl, defined as a "high-density urban settlement" promoting mixed land use, public transportation, more efficient use of energy and other resources, greater accessibility, land preservation for agriculture and vegetation, as well as reuse of infrastructures that are already existing rather than building new ones. Such multiple land use also ensures healthier communities by closer proximity of work, living and recreational spaces and embracing activities like cycling and walking (Abdullahi et al., 2015).

Gothenburg and Helsingborg are the two Swedish cities that are excellent examples of model compact city (Figure 1.5). These cities are undergoing population expansion and hence have been focussing on creating dense developed areas with public transport, green spaces as well as a combination of commercial, residential and recreational spaces to enhance accessibility (Bibri et al., 2020). Portland, USA follows policies that favour compact neighbourhoods and public transit systems. The city limits its urban sprawl through its Urban Growth Boundary policy, hence preserving natural and agricultural lands while supporting urban density (Raghubanshi, 2024).

The goal should be to improve land use planning such that urban development does not affect the land required for agriculture and biodiversity and turning low-density cities into high-density cities.

1.4.2 Renewable Energy and Energy Efficiency

This principle focuses on efficient energy use, modified systems of energy supply and harnessing renewable energy to minimize or replace the use of fossil fuel-based

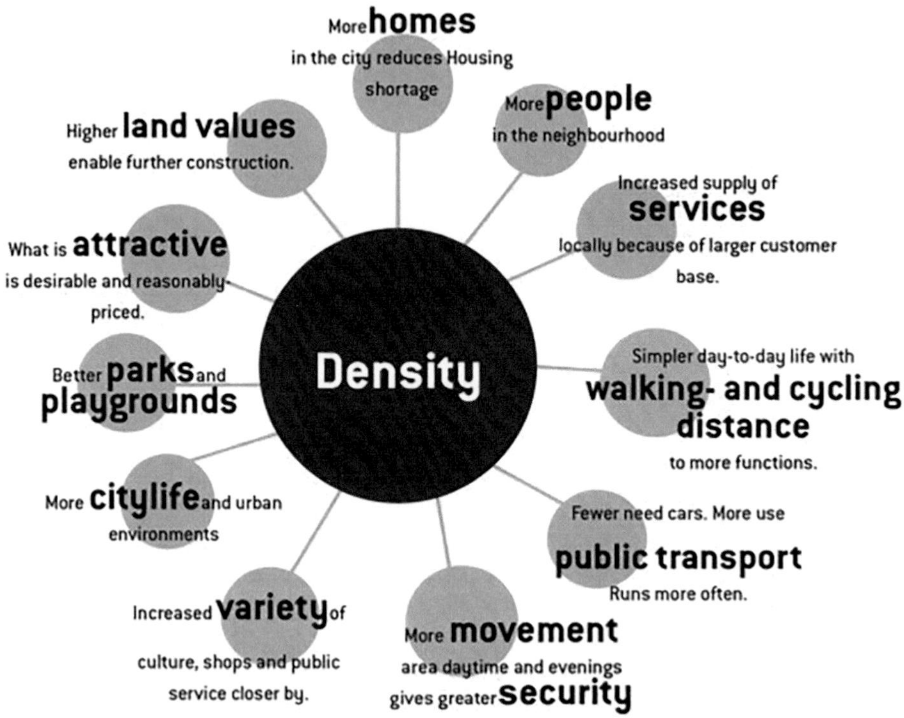

FIGURE 1.5 Benefits of a High-Density City (reproduced from Bibri et al., 2020, CC BY 4.0.).

energy. The city should strive to become the producer of energy instead of being its consumer. Renewable energy sources must be identified to execute the plan along with an assessment of the availability and cost of technology, involving a suitable combination of the sources which can include solar, wind, mini-hydro, biomass energy projects based on the city location and characteristics. Sources available can vary as per different locations and micro-climate of the city for which a distributed or decentralized energy supply system is required that enables the utilization of locally produced renewable energies.

Conventionally, energy supply has been via large centralized systems in cities which constitutes extensive transport and distribution of power through long distances causing power wastage along the way. A renewable city, hence, would showcase a decentralized system (Figure 1.6) where energy production is local and closer to target consumers leading to benefits such as decreased power loss, greater control in energy production and enhanced resilience towards disasters. This approach has been adopted in various cities today such as Amsterdam, Netherlands which promotes community-based energy production and consumption through a decentralized energy framework using smart grids (Swain et al., 2021); Berlin, Germany which promotes local energy trading and enhanced grid stability (Rozhkov, 2024) and Melbourne, Australia which has a DE model promoting renewable energy use as well as local energy markets (Adil & Ko, 2016).

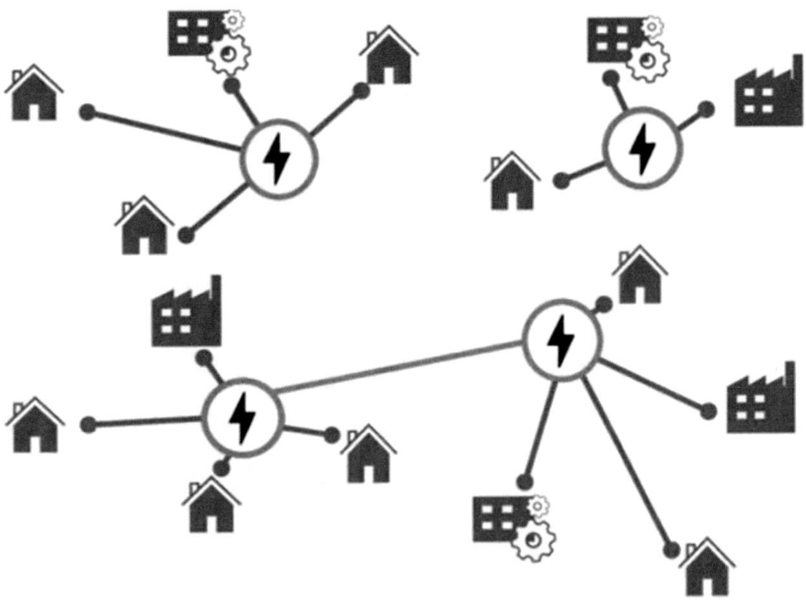

FIGURE 1.6 Decentralized energy distribution model (reproduced from Vezzloi et al. 2018 , CC BY 4.0).

Another goal is to replace half of a city's energy generation through coal-based power stations with renewable energy sources to provide a meaningful contribution in tackling issues like climate change, carbon emissions, public health and so on. On an individual level, buildings can be equipped with solar panels as well as smart technology that shuts down energy consuming appliances in the absence of users. Transport can be another important area to apply this principle by switching to electric power derived from sunlight, moving towards the objective of carbon-neutrality.

1.4.3 Water Management

Proper water management in an urban area would cover aspects including reduced consumption of water, restoration of existing and development of new water resources, good water quality, efficient water conservation systems designed for rain or storm water harvesting and groundwater sequestration, along with ensuring awareness among citizens in conservational practices such as efficient use and collection of water, wastewater reuse and recycling.

Drainage systems of the city need to be effective and working during excess rain or storm to regulate urban flooding. There must be proper treatment of wastewater and such facilities may use nature-derived solutions such as algae, to filter the grey water and improve its quality, further allowing recycling or safe disposal in natural water resources. At the household level, rainwater harvesting system needs to be installed to utilize collected water in daily activities, along with appliances or technology that allows controlled use of water such as low-flush toilets. For agriculture, drought-resistant crops need to be used that can thrive on less water.

Like the energy supply system, the water supply also needs to be decentralized so as to reduce energy consumption and water loss as well as to retain the quality of water. Safe, clean water is essential for public health. For instance, it is mandatory in new urban settlements of Almere, Netherlands to have a specific part of land dedicated to open water bodies which aids in conservation of rainwater, reduces flood risk and improves freshwater resource availability. In contrast, Hsingchu, Taiwan is an example of a city that faces seasonal droughts and floods and exerts significant pressure on its freshwater resources due to urbanization. For such a city, a decentralized rainwater management system is essential to mitigate its water-related issues (Schuetze & Chelleri, 2013).

1.4.4 Green Buildings

Lehmann suggests essential aspects of the principle including designs that cause zero emissions and utilize less energy, introducing renewable energy sources such as solar panels in architecture, retrofitting of existing buildings to make them more sustainable, passive designs that use natural resources to regulate a building's environment via ventilation, daylight, vegetation and enhance efficient use of energy. The goal is to consume minimal resources as well as materials. Buildings need to have high longevity, should consume less and generate greater energy as well as collect and treat their water.

Limiting urban sprawl by supporting compact urban planning is another important area to consider. As shown in Figure 1.7, this can be done by developing high-density areas by efficient use of land and building vertical infrastructures instead of horizontal ones and further incorporation of vegetation in them (Artmann et al., 2017). Materials used for construction purposes should be locally derived, requiring less transport and recyclable at the end of the building's life cycle. Such materials can be manufactured at a local scale that are environment-friendly and lead to reduced energy consumption, hence also supporting the urban economy.

1.4.5 Urban Ecology and Biodiversity

The principle focuses on conservation, restoration and rehabilitation of areas of biodiversity along with quicker access of the urban population to green spaces for relaxation and recreation, promoting a healthy community. City densification should go hand in hand with preserving land for creating open spaces of vegetation for supporting the urban wildlife, reducing the UHI or "urban heat island" effect, temperature regulation and purification of air. Other objectives include necessitating tree belts wherever possible, maintaining the present ecosystems and involving urban planting in every developmental project, hence propagating carbon sinks in the city. Natural resources need to be conserved and efficiently used every step of the way.

Green infrastructure refers to the integration of natural systems such as green spaces into urban infrastructure which is possible via designs such as green roofs, wall plantations, street tree belts, gardens, parks, along with inclusion of water bodies, urban forests, agricultural land and so on. Planning green infrastructure closer to urban dwellings is beneficial. It not only supports the urban ecology and

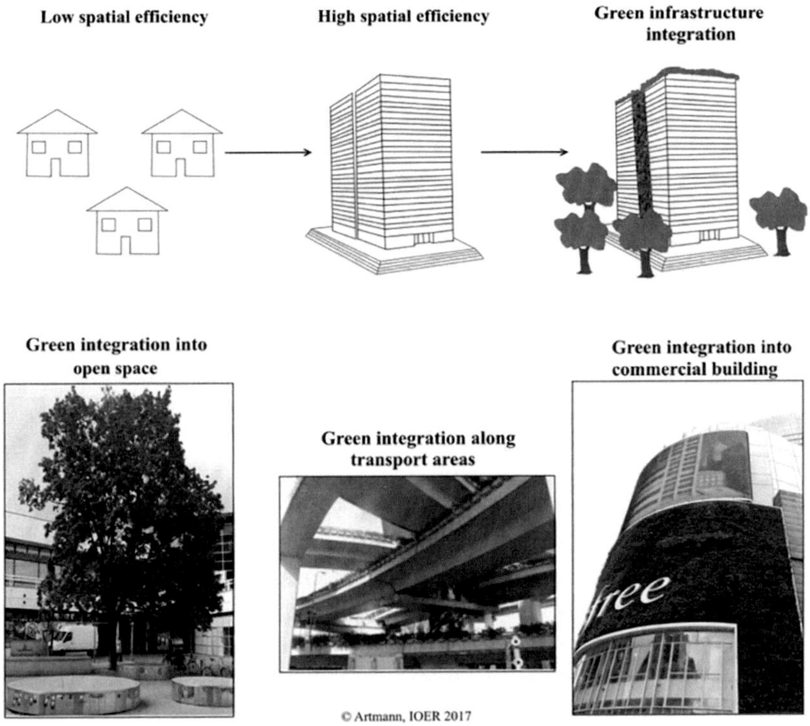

FIGURE 1.7 Spatially efficient green integration (Artmann et al. 2017).

biodiversity but also the local economy by increasing the flux of tourists or recreational activities in aestheticized areas (Artmann et al., 2017). Singapore is known as "City in a Garden" as it incorporates green infrastructure such as vertical gardens and green roofs in residential and commercial buildings, promoting biodiversity and reducing urban heat despite being highly populated. Curtiba in Brazil comprises of extensive green spaces, corridors and parks combined with urban design (Raghubanshi, 2024).

Urban Ecosystem Services include all the benefits provided by the urban ecosystems to the residents whether in terms of economy or quality of life, ranging from climate regulation via tree-shading and evapotranspiration processes, biodiversity including valuable species thriving in urban natural areas that act as habitats, pollination services, enhanced public physical and mental health to carbon intake by urban trees and soils. These services can only be enhanced with the enhancement of urban ecosystems (Breuste et al., 2013).

Soil sealing refers to replacement of the vegetation cover over soil with materials that are less or non-permeable for developmental purposes such as building roads, sidewalks, pavements, and so on. This activity is characteristic of rising urbanization and adversely impacts ecosystem and its services causing destruction of soil biodiversity, changes in the micro-climate of the area, reduction in natural habitats, increased surface runoff and reduction in groundwater recharge. Soil sealing must be avoided wherever possible, along with supporting urban planning with open natural areas, ensuring infiltration of water and using materials that are permeable (Andersson et al., 2014).

In conclusion to all the aspects above, the main goal here should be increasing the green area percentage, combined with urban densification projects.

1.4.6 Sustainable Transportation

Sustainable transportation not only includes a reliable public transport system but also restructuring the whole city such that greater opportunities can be created for the use of public transport services. A dense city will provide for easier access to destinations using public transport while a low-density city will create dependence on personal automobiles such as cars to reach to places in time. This calls for careful urban planning involving compact city design, aesthetic public spaces, streets designed for transport alternatives like cycling and walking, greater public space connectivity, vehicles running on renewable energy, as well as a working combination of forms of public transport like bus, metro, and rail. Designs such as wider roads and freeways are unsustainable as the former leads to increase in traffic and carbon emissions, and eats away the land for green spaces, whereas the latter generates dependence on cars as well as promotes urban sprawl.

Nowadays we witness modern electric-rail systems across important metropolitan cities such as Delhi in India (Figure 1.8) and Beijing in China that are needed for managing traffic, enhanced public space network, reducing usage of fossil fuels as well as automobiles, and hence progressing towards carbon-neutrality.

Transit-oriented development (TOD) is an important strategy of urban planning which focuses on developing compact urban communities around stations of public transport, in other words, creating residential, recreational and business spaces that are walking distance apart from public transportation (Figure 1.9). TODs reduce urban ecological footprint by reducing utilization of private automobiles, land preservation for green spaces as well as healthier dense communities that have easy access to public spaces (Ibraeva et al., 2020). Urban restructuring around public transit is the future.

Curitiba, Brazil has a Bus Rapid Transit system that transports citizens efficiently, combatting traffic congestion as well as air pollution. Copenhagen, Denmark has made cycling as its primary mode of transportation which heavily reduces dependency on fossil fuels (Raghubanshi, 2024).

1.4.7 Waste Management

Sustainable waste management calls for a closed-loop or circular system wherein waste produced is utilized as a resource. The goal should be zero-waste by using strategies like reusing, reducing, recycling as well as composting, with application of waste-to-energy technology, and special focus on treatment of industrial and electronic waste. Inspiration should be taken from an ecosystem's self-sustaining nature in which there is a cyclic and balanced flow of nutrients and resources, with no wastage. Such an "eco-efficient city", as Newman (2010) states, has lower ecological

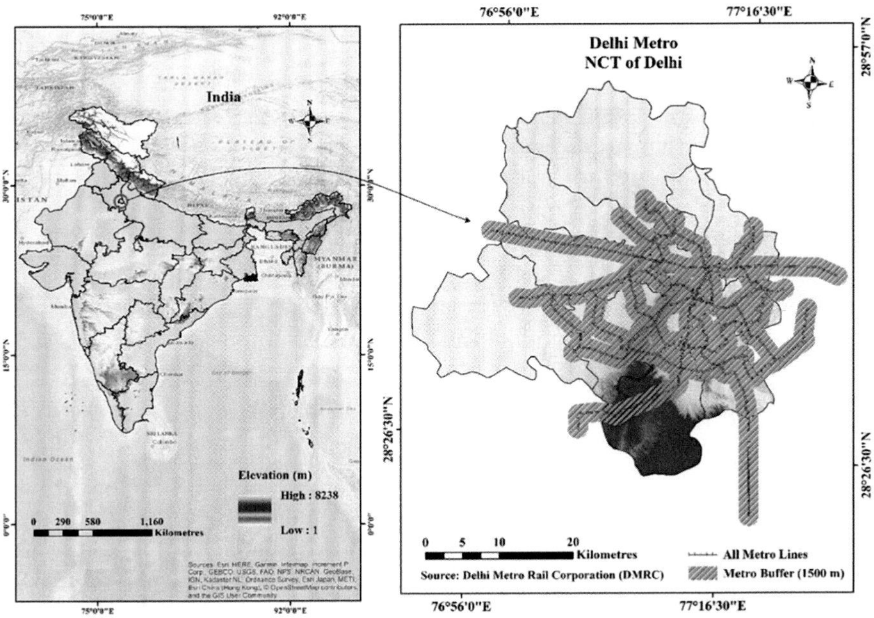

FIGURE 1.8 Delhi metro railway network (reproduced from Begam et al. 2024, CC BY 4.0).

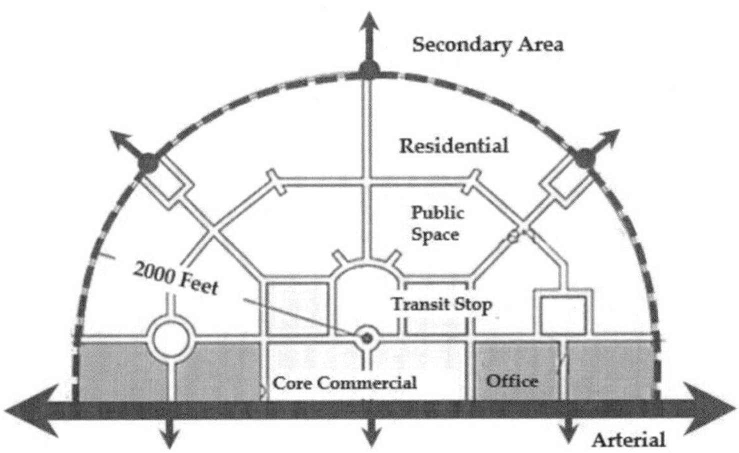

FIGURE 1.9 TOD model (reproduced from Nafi et al. 2021, CC BY 4.0).

footprint due to reduced resource utilization as well as waste production. A city needs to be perceived as an ecosystem in itself with its complex workings and cycles where what has been viewed as "waste" is re-envisioned as a valuable input in other urban requirements, such as energy production, building materials, and so on rather than a linear pipeline model that involves overexploitation of depleting resources and increased volume of waste. Other strategies include waste prevention, extended producer responsibility, building centres for recycling and remanufacturing, nature-derived recyclable goods and materials.

The application of this principle requires strong inclusion of the *eco-efficiency agenda* in urban planning that is based on the *cradle-to-cradle* concept by McDonaugh and Braungart (2002). This concept considers resources as "nutrients" that are everlasting in any form and can be re-introduced into the economy.

We can find one such example in the urban locality of Stockholm, named Hammarby Sjöstad where stoves utilize biogas produced from community generated wastewater (Newman, 2010). Many such instances can be observed today as well as created by shifting to resource recovery and circular supply chains for a sustainable economy, reduced overall consumption and healthier environment.

1.4.8 Community, Quality of Life and Social Equity

GU aims to create conditions for a healthy, fulfilling and a high-quality life, enabling people to make sustainable lifestyle choices such as growing their own food, having access to environment-friendly goods, walking or cycling, not needing a car and so on (Beatley, 2012). Important aspects of this principle are mentioned below:

1.4.8.1 Social Inclusion and Affordable Housing

Focus should be on creating mixed-use compact neighbourhoods with affordable housing, places of recreation, nature, business, education, culture, and so on close to each other which reduces traffic and pollution. The housing facility should be economically viable and inclusive of people regardless of age, profession, status and so on. Such a city is highly liveable. Seattle is a fine example of a city that prioritizes social equity as opinions and concerns of marginalized communities especially are taken into serious consideration in drafting environmental plans (USGBC, 2025).

1.4.8.2 Local Economy, Culture and Sense of Place

Development of local economy should be emphasized upon for greater sustainability. Creating local job opportunities and supporting local businesses and communities empowers the city population, reduces the need to migrate or travel far for jobs, increases quality of life and is correlated with reduced ecological footprint. It goes hand in hand with preserving the city's unique identity and culture, utilizing it to develop social capital which in turn strengthens the urban economy. The creativity and heritage of the city must be respected and further enhanced, forming a strong "sense of place".

1.4.8.3 Local Food Supply

Local production and supply of food should be encouraged via urban farming and slow food practices. Keeping sufficient land preserved for agriculture should be an important component of urban planning, along with communal spaces for growing fruits and vegetables, roof gardens, encouraging composting using local organic waste. This will help cut down on transportation as well as food price, promote healthier diets as well as better waste management. The goal should be to produce 50% of the urban food organically without any artificial chemicals.

1.4.8.4 Educational Reforms

The concept of GU and sustainability (whether in design, architecture, daily life and behaviour) needs to be included in early education and other areas of knowledge for bringing change in people's attitude as well as further dissemination of sustainable ideas and creation of green jobs.

1.4.9 Policy and Governance

To execute the principles and objectives of GU in existing cities, effective urban governance and policy is necessary.

Sareen and Waagsaether (2023) identify three crucial shifts: inclusion, alignment and orientation in governance that are required for smooth urban transition to sustainability. *Inclusion* refers to the involvement of marginalized groups in decision-making processes. *Alignment* means that policies are well-aligned across different governmental sectors, that is, policies at regional, state and national levels support each other to fulfil common sustainability goals. *Orientation* refers to a shift from the interests of powerful individuals towards inclusive low-carbon urban scenario.

Socio-political issues can arise from poor urban governance and prioritizing short-term economic and political interests (rather than long-term ecological goals) that can act as barriers to GU. One such case is of *Shenzhen*, first low-carbon eco-city of China. It became China's first Special Economic Zone in 1980s, and in 1990s it further expanded, overtaking the surrounding rural land which led to incessant urbanization, rising migrant population, urban sprawl and lower productivity. There began a huge shortage of land, poverty hike and degrading environmental conditions. Since 2000s, the city has been involved in developing low-carbon eco-districts and refurbishing brownfield sites to address land shortage and achieve higher economic growth. It is argued that the adoption of GU practices here may be for the wrong reasons and limited to boosting land supply and value as it has failed to protect the rights of the displaced rural and urban village population (Ng, 2019).

In Arab cities, state-led governance is observed which prioritizes social stability and scraps any sustainable plans. There is also a lack of community engagement that prevents implementation of such initiatives (Verdeil, 2019). In few cities of Netherlands, a gap is observed between community and rigid municipality that halters citizen participation in green initiatives (Romero-Muñoz et al., 2024). Other common challenges include insufficient funding for maintenance of green spaces, lack of public awareness, exclusion of certain groups in decision-making related to green ventures, overlooking existing social injustices and failing to engage communities to work together (Liette, 2014).

GU addresses these issues by emphasizing collaboration among various stakeholders including governments, private sector, NGOs, and so on to overcome economic and political constraints. Green governance calls for sustainability measures integrated in policies and decisions, for them to be properly reinforced and maintained over the city, which can include mixed land use policies, public awareness, densification programs, financing low-carbon technologies, incentivizing sustainable actions and so on. To follow the laid-out strategies and direction, adequate government support is required. Urban governmental authorities need to first follow sustainable practices and bring change in their decision-making. Since this is a two-way street, citizens need to be aware and actively participate in forming a sustainable community as well as putting forward their concerns to the authorities (Lehmann, 2011).

There are certain places where sustainable development and economic growth co-exist. Nordic countries, France, Germany, Netherlands follow green governance model involving green investments, circular economy, supportive policies and laws, and inclusive development (Lisenko, 2024). Ponorogo, Indonesia showcases effective governance in managing open green spaces through awareness and coordination among government, NGOs, private sector and the community leading to proper care of these spaces (Wiratama et al., 2023). *Unilever*, a prominent global consumer goods company, is an example of green corporate governance which has implemented various sustainable practices such as setting measurable goals for reducing water consumption, greenhouse gas emissions and waste production. It also promoted stakeholder engagement and transparency, worked on sourcing raw materials sustainably and limiting packaging waste (Ijomah et al., 2024).

1.5 FUTURE TRENDS

With rising technological interventions and changing priorities globally with time, urbanization today is being moulded accordingly which can be mainly observed in the form of the smart city project as well as growing climate resilience, discussed below.

1.5.1 Smart Cities

A smart city is the integration of modern technological advances including Information Communication Technology (ICT), Internet of Things (IoT), artificial intelligence (AI), data analytics and so on with urban design to produce smart solutions towards urban sustainability and enhancing quality of life. A smart city project carves a pathway in building an environmentally, economically and socially sustainable urban community that is growth-oriented and inclusive. It focuses on infusing technology in all aspects of daily urban life including infrastructure, healthcare, transportation, energy, innovation and public safety. Such a city is greener, faster, safer and friendlier (Mohanty et al., 2016; Romanelli, 2021).

Okello and Akoko (2023) outline important emerging trends that are shaping today's smart cities, mentioned below:

- ***Smart Energy*:** features efficient use of energy, sustainable energy infrastructure including "smart grids" for energy tracking and regulation, supporting policies and reduced emissions as well as climate change impact.
- ***Smart Mobility*:** includes sustainable transport systems and solutions like zero emission vehicles, electrification, smart management of traffic, carpooling, and so on for minimizing noise pollution, transportation costs, traffic congestion and increasing speed as well as safety.

- ***E-governance*:** strives towards efficient and transparent delivery of government and public services through the use of information technology. It can include e-services like online voting, license application or digital passports, with benefits like revenue savings, positive relationship between the people and government via free flow of services and collaboration between the two.
- ***Smart Healthcare*:** focuses on affordable and accessible healthcare, high-quality services, better life quality and better engagement between healthcare providers and patients.

Smart cities can be very helpful in strengthening climate resilience by utilizing innovative technologies to reduce carbon footprints and increase energy efficiency. They can analyse available data on climate and disaster risks to provide climate resilience strategies, useful under climatic stress. Such a city is also characterized by the use of renewable energy sources that reduce the dependency on fossil fuels, aligning with principles of GU (Mallick, 2024; Ruiz-Vanoye et al., 2024; Sharma et al., 2024).

There are several budding examples of smart cities today, including in India. Singapore, Barcelona, Stockholm and Spain are leading smart cities, utilizing IoT and smart grids in urban management. China is working towards developing a smart batch of 90 cities, districts and towns. A $4.2 billion smart city project in Vietnam, near Hanoi is undergoing with expected completion in 2028. Indian Government launched its *Smart Cities Mission* (SCM) in 2015, planning to develop 100 smart cities with enhanced and sustainable living, infrastructure, environment, economy, transport and governance. Cities like Bhopal, Chennai, Pune, Ahmedabad, with Nagpur being one of the top-performing smart cities, are involved in the mission with many others (Gupta & Hall, 2017; Höffken & Limmer, 2019; Kandpal, 2018; Okello & Akoko, 2023).

Although technology enhances urban living, it can also pose a few serious threats for the same. Smart cities require data collection from citizens which can be done without their full awareness of how the data will be used or protected. Hence it can easily lead to data breaches and privacy violations. Differences in technological accessibility can create a digital divide, acting as a disadvantage for those citizens who lack digital skills or resources which can further lead to their exclusion from the benefits of other innovations and smart city services. This digital disparity can in turn deepen the social inequalities and marginalization (Hysa et al., 2024). To tackle this, sincere efforts in spreading digital literacy are needed.

1.5.2 Climate Resilience

Resilience refers to the ability of a city to withstand extreme events and shocks, preventing destructive damage and loss to its structure and inhabitants. With increasing instances and intensity of natural disasters in lieu of climate change today, cities are more vulnerable than ever to their negative impact, not excluding risks from human-induced calamities (Asprone & Manfredi, 2015). For a city to remain resilient in these times, integrating DRM in urban planning has become a crucial goal globally and has also been incorporated in international agreements such as the Paris

Agreement, the 2030 Agenda for Sustainable Development and Sendai Framework for Disaster Risk Reduction (Raikes et al., 2022). DRM has to be an integral part of the urban designing process, including development policies, land use plans, infrastructure design and building codes (Benson, 2016), as well as introducing appropriate mitigation and adaptation measures.

Implementation of such climate resilience measures is being observed in urban cities including development of green infrastructure which is required for carbon sequestration and micro-climate regulation and white roofs that enhance solar reflectivity and reduce heat flux (Rafael et al., 2016). Other focus areas include land use that is climate sensitive, water conservation, improvement of drainage systems, emergency planning, disaster awareness and capacity-building, early-warning systems, resilient transport and housing, healthcare institutions that are responsive, maintenance of ecosystem services, climate relief funding to vulnerable communities and their protection (Brown et al., 2012).

Hazard risk assessment is now a necessity in construction or developmental projects utilizing technology in identifying areas of high-risk and enhancing urban resilience accordingly (Rezvani et al., 2023). *Nature-based solutions* (NbS) harness the ability of nature to benefit society, economy and biodiversity, involving creation of more ecosystem services, natural resource management, restoration and protection of ecosystems and so on. It is widely reported that NbS are effective in minimizing climate-change impact as well as hydrometeorological risks globally (Turner et al., 2022).

Despite the present economic and political challenges, climate resilience needs to be infused to the core of urban planning ensuring that the cities are not just smart and green but also can withstand as well as recover quickly from any kind of calamity (Shahrabani & Apanavičienė, 2022).

1.6 CONCLUSION

GU has emerged as a critical response to the rising need for sustainable urbanization. It integrates environmental, economic and social dimensions to create holistic urban spaces that minimize carbon footprints, strengthen biodiversity and advocate sustainable living. As cities struggle with overpopulation, resource depletion and climate change, GU paves the way towards resilience. However, the time to act is now, before human actions push the earth beyond its ecological limit. Policy-makers, urban planners and citizens must collaborate and prioritize environmental harmony over consumerism, incorporating the principles of GU in urban planning, developmental reforms and lifestyle.

Along with the purpose and detailed principles, this chapter explored the historical evolution of GU as well as the future trends that continue to modify the concept with growing innovation, technological advancements, smart city projects, inclusive governance and strategies for climate resilience.

Further important areas of research involve:

- Addressing the context-specific challenges faced by cities of developing countries in adopting to GU strategies

- Analysing the effectiveness of the green strategies in bringing the desired objectives to reality
- Exploring technological advancements in green infrastructure
- Formulating policy frameworks that align with GU

REFERENCES

Abdullahi, S., Pradhan, B., Mansor, S., & Shariff, A. R. M. (2015). GIS-based modeling for the spatial measurement and evaluation of mixed land use development for a compact city. *GIScience & Remote Sensing*, 52(1), 18–39. https://doi.org/10.1080/15481603.2014.993854

Abubakar, I. R., & Dano, U. L. (2020). Sustainable urban planning strategies for mitigating climate change in Saudi Arabia. *Environment, Development and Sustainability*, 22(6), 5129–5152. https://doi.org/10.1007/s10668-019-00417-1

Adil, A. M., & Ko, Y. (2016). Socio-technical evolution of decentralized energy systems: A critical review and implications for urban planning and policy. *Renewable and Sustainable Energy Reviews*, 57, 1025–1037. https://doi.org/10.1016/j.rser.2015.12.079

Andersson, E., Barthel, S., Borgström, S., Colding, J., Elmqvist, T., Folke, C., & Gren, Å. (2014). Reconnecting cities to the biosphere: Stewardship of green infrastructure and urban ecosystem services. *Ambio*, 43(4), 445–453. https://doi.org/10.1007/s13280-014-0506-y

Artmann, M., Kohler, M., Meinel, G., Gan, J., & Ioja, I. C. (2017). How smart growth and green infrastructure can mutually support each other: A conceptual framework for compact and green cities. *Ecological Indicators*, 96, 10–22. https://doi.org/10.1016/j.ecolind.2017.07.001

Asprone, D., & Manfredi, G. (2015). Linking disaster resilience and urban sustainability: A glocal approach for future cities. *Disasters*, *39*(s1). https://doi.org/10.1111/disa.12106

Barton, J. R. (2019). Green Urbanism. In Anthony M. Orum (Ed.), *The Wiley Blackwell encyclopedia of urban and regional studies* (pp. 1–6). Wiley. https://doi.org/10.1002/9781118568446.eurs0127

Beatley, T. (2000). *Green urbanism: Learning from European cities*. Island Press.

Beatley, T. (2012). *Green cities of Europe* (T. Beatley, Ed.). Island Press/Center for Resource Economics. https://doi.org/10.5822/978-1-61091-175-7

Beatley, T., & Newman, P. (2013). Biophilic cities are sustainable, resilient cities. *Sustainability*, 5(8), 3328–3345. https://doi.org/10.3390/su5083328

Begam, S., Jha, P., Yadav, P. K., Joy, M. S., Rawat, P., & Bansal, T. (2024). Urban transformation through transit: The case of Delhi metro. *Discover Cities*, 1(1), 31. https://doi.org/10.1007/s44327-024-00035-1

Benson, C. (2016). *Promoting Sustainable Development through Disaster Risk Management*. Asian Development Bank Sustainable Development Working paper Series, number 41. https://www.adb.org/sites/default/files/publication/182652/sdwp-041.pdf

Bhardwaj, K. (2024). The historical sustainable cities in India: A survey of sustainable practices followed in cities of India in ancient and medieval times. *IOSR Journal of Humanities and Social Science*, 29(9), 31–34. https://doi.org/10.9790/0837-2909073134

Bhattacharya, Dr. S. (2021). Smart city a vision of India to provide people with better lifestyle and move towards a sustainable development. *International Journal for Research in Applied Science and Engineering Technology*, 9(11), 1050–1054. https://doi.org/10.22214/ijraset.2021.38972

Bibri, S. E., Krogstie, J., & Kärrholm, M. (2020). Compact city planning and development: Emerging practices and strategies for achieving the goals of sustainability. *Developments in the Built Environment*, *4*, 100021. https://doi.org/10.1016/j.dibe.2020.100021

Bignami, D. F. (2020). Disaster risk assessment, reduction and resilience: Their reciprocal contribution with urban planning to advance sustainability. In G. Dall'O' (Ed.), *Green planning for cities and communities: Novel incisive approaches to sustainability* (pp. 67–94). Springer International Publishing.

Breuste, J., Haase, D., & Elmqvist, T. (2013). Urban landscapes and ecosystem services. In S. Wratten, H. Sandhu, R. Cullen, & R. Costanza (Eds.), Ecosystem services in agricultural and urban landscapes (pp. 83–104). Chichester, UK: John Wiley & Sons, Ltd.

Brown, A., Dayal, A., & Rumbaitis Del Rio, C. (2012). From practice to theory: Emerging lessons from Asia for building urban climate change resilience. *Environment and Urbanization*, 24(2), 531–556. https://doi.org/10.1177/0956247812456490

Brundtland, G. H., & Khalid, M. (1987). *World commission on environment and development: Our common future*. Oxford University Press.

Carson, R. (1962). *Silent spring*. Houghton Mifflin.

C/O City. (2019, August 13). *Urban ecosystem services – C/O City*. https://www.cocity.se/om-oss/urban-ecosystem-services/

Cucoranu, C. V., & Popescu, D. (2023). *Sustainable urban development: Importance and challenges*. In A. I. Vodă, A. M. Bercu, & L.-D. Radu (Eds.), Proceedings of the International Conference on Sustainable Development of European Smart Cities (SmartEU 2023) (pp. 213–224). Iași, Romania: Alexandru Ioan Cuza University Press.

Dholavira. (2006). Wikimedia Commons. https://en.wikipedia.org/wiki/File:Dholavira1.JPG

Ehrlich, P. (1968). *The Population Bomb*. Ballantine Books.

Farinha-Marques, P., Lameiras, J. M., Fernandes, C., Silva, S., & Guilherme, F. (2011). Urban biodiversity: A review of current concepts and contributions to multidisciplinary approaches. *Innovation: The European Journal of Social Science Research*, 24(3), 247–271. https://doi.org/10.1080/13511610.2011.592062

Farr, D. (2008). *Sustainable urbanism: Urban design with nature*. Wiley.

Gómez-Villarino, M. T., Gómez Villarino, M., & Ruiz-Garcia, L. (2020). Implementation of urban green infrastructures in peri-urban areas: A case study of climate change mitigation in Madrid. *Agronomy*, 11(1), 31. https://doi.org/10.3390/agronomy11010031

Green Rating for Integrated Habitat Assessment. (2021). https://www.grihaindia.org/

Gupta, K., & Hall, R. P. (2017). The Indian perspective of smart cities. *2017 Smart City Symposium Prague (SCSP)*, 1–6. https://doi.org/10.1109/SCSP.2017.7973837

Gürçam, S. (2024). Paving the way for climate resilience through sustainable urbanization: A comparative study. *Lectio Socialis*, 8(1), 17–34. https://doi.org/10.47478/lectio.1396483

Höffken, J. I., & Limmer, A. (2019). Smart and eco-cities in India and China. *Local Environment*, 24(7), 646–661. https://doi.org/10.1080/13549839.2019.1628730

Howard, E. (1902). *Garden cities of tomorrow*. ATC Books/Faber and Faber.

Hysa, E., Kruja, A., Rehman, N. U., & Laurenti, R. (2020). Circular economy innovation and environmental sustainability impact on economic growth: An integrated model for sustainable development. *Sustainability*, 12(12), 4831. https://doi.org/10.3390/su12124831

Hysa, X., Guazzo, G. M., Çekani, V., & Rosati, P. (2024). Privacy and smart cities: A bibliometric analysis. In A. Visvizi, O. Troisi, & V. Corvello (Eds.), *Research and Innovation Forum 2023. RIIFORUM 2023. Springer Proceedings in Complexity* (pp. 151–161). Springer. https://doi.org/10.1007/978-3-031-44721-1_12

Ibraeva, A., Correia, G. H. de A., Silva, C., & Antunes, A. P. (2020). Transit-oriented development: A review of research achievements and challenges. *Transportation Research Part A: Policy and Practice*, 132, 110–130. https://doi.org/10.1016/j.tra.2019.10.018

Ijomah, T. I., Nwabekee, U. S., Agu, E. E., & Abdul-Azeez, O. Y. (2024). The evolution of environmental responsibility in corporate governance: Case studies and lessons learned. *International Journal of Frontline Research in Science and Technology*, 3(2), 019–037. https://doi.org/10.56355/ijfrst.2024.3.2.0045

Imam, A. U. K., & Banerjee, U. K. (2016). Urbanisation and greening of Indian cities: Problems, practices, and policies. *Ambio*, 45(4), 442–457. https://doi.org/10.1007/s13280-015-0763-4

Jenks, M., & Burgess, R. (2002). *Compact cities* (R. Burgess & M. Jenks, Eds.). Routledge. https://doi.org/10.4324/9780203478622

Kandpal, V. (2018). A case study on smart city projects in India. *Companion of the The Web Conference 2018 on The Web Conference 2018 – WWW '18*, 935–941. https://doi.org/10.1145/3184558.3191522

Karlenzig, W., Marquardt, F., White, P., Yaseen, P., & Young, R. (2007). *How green is your city, the sustainLane US city rankings.* New Society Publishers.

Khamdamov, S.-J., & Usmanov, A. (2024). Sustainable cities and communities: urban planning and development strategies. *European Journal of Artificial Intelligence and Digital Economy*, 1(7), 76–81. https://doi.org/10.61796/jaide.v1i7.794

Kineman, J. J., & Anand, D. (2015). *Roots of sustainability in ancient India. Proceedings of the 58th Annual Meeting of the ISSS – 2014 United States*, *1*(1). https://journals.isss.org/index.php/proceedings58th/article/view/2349

Koller, M., Eckert, K., Ferber, U., Gräbe, G., Verbücheln, M., & Wendler, K. (2022). Resource management as part of sustainable urban district development. *Sustainability*, 14(7), 4224. https://doi.org/10.3390/su14074224

Konstantinou, P., Stathakis, G., Mountzouri, A., & Nomikou, M-G. (2020). Technology and social inequality within smart cities. East–West Journal of Economics and Business, 23(2), 51–70. https://www.u-picardie.fr/eastwest/stat_doc.php?doc=249&utm_source=chatgpt.com

Kowarik, I., Fischer, L. K., & Kendal, D. (2020). Biodiversity conservation and sustainable urban development. *Sustainability*, 12(12), 4964. https://doi.org/10.3390/su12124964

Kumarappa, J. C. (1945). *Economy of Permanence a quest for a social order based on non-violence.* Sarva Seva Sangh Prakashan. www.mkgandhi.org

Lehmann, S. (2010). Transforming the city for sustainability: The principles of green urbanism. *Journal of Green Building*, 6(1), 104–113. https://doi.org/10.3992/jgb.6.1.104

Lehmann, S. (2011). What is green urbanism? Holistic principles to transform cities for sustainability. In J. A. Blanco&radmand (Eds.),*Cmate change — research and technology for adaptation and mitigation* (p. 243). InTech. https://doi.org/10.5772/23957

Lehmann, S., & Mainguy, G. (2010). Green urbanism: Formulating a series of holistic principles. *SAPI EN. S. Surveys and Perspectives Integrating Environment and Society*, *3*(2). https://journals.openedition.org/sapiens/1057#article-1057

Liette, G. (2014). Social justice and the "green" city. *URBE — Revista Brasileira de Gestão Urbana*, 6(541), 158. https://doi.org/10.7213/urbe.06.002.SE01

Lisenko, V. (2024). Sustainable economy and effective environmental governance: Key principles and global best practices. *Revista Moldovenească de Drept Internaţional Şi Relaţii Internaţionale*, 19(2), 52–64. https://doi.org/10.61753/1857-1999/2345-1963/2023.19-2.04

Mallick, R. (2024). *Smart cities and climate resilience* (pp. 401–412). https://doi.org/10.4018/979-8-3693-6522-9.ch021

Mangweta, R., Mokoele, N. J., & Monama, S. A. (2022). Building sustainable cities to address urban sprawl: A reflective analysis towards achieving SDGs. *EUREKA: Social and Humanities*, 6, 72–78. https://doi.org/10.4018/979-8-3693-6522-9.ch021

Marsh, G. P. (1865). *Man and nature; or, physical geography as modified by human action.* Charles Scribner.

Matan, A., & Newman, P. (2015). Green urbanism in the Indian Ocean region. *Journal of the Indian Ocean Region*, 11(1), 60–73. https://doi.org/10.1080/19480881.2015.1017228

McDonaugh, W., & Braungart, D. (2002). *Cradle to cradle: Remaking the way we make things.* North Point Press.

Mcfarlane, I., Alyanak, L., Jensen, J., Kollodge, R., Daldin, J., Jayaram, T., Ratcliffe, L., Trautwein, C., Baker, D., Botev, N., Garbett, A., Gietel-Basten, S., Luchsinger, G., Nandagiri, R., Sear, R., Sobotka, T., Armitage, A., Chalasani, S., Eschenbaecher, J.-H., … & Snow, R. (2023). *State of World Population report 2023*. www.unfpa.org/swp2023/YouGovData

McHarg, I. L. (1969). *Design with nature*. Nature History Press.

Meadows, D. H., Randers, J., Meadows, D. L., & Behrens, W. W. (1972). *The limits to growth: A report for the club of rome's project on the predicament of mankind*. Universe Books.

Mohanty, S. P., Choppali, U., & Kougianos, E. (2016). Everything you wanted to know about smart cities: The Internet of things is the backbone. *IEEE Consumer Electronics Magazine*, 5(3), 60–70. https://doi.org/10.1109/MCE.2016.2556879

Mudaliar, P., & Kashwan, P. (2024). Poverty and pollution. *Environment: Science and Policy for Sustainable Development*, 66(1), 7–18. https://doi.org/10.1080/00139157.2023.2269041

Muir, J. (1911). *My first summer in the Sierra*. Houghton Mifflin.

Nabila, N. (2021). The concept of garden city and its relevancy in modern city planning. *Southeast University Journal of Architecture*, *1*(1), 1–7. www.seu.edu.bd/seuja

Nafi, S., Furlan, R., Grosvald, M., Al-Matwi, R., & Lifam Marthya, K. (2021). Transit-oriented development in Doha: The case of the Al Sadd neighborhood and Hamad Hospital metro station. *Designs*, 5(4), 61. https://doi.org/10.3390/designs5040061

Newman, P. (2010). Green urbanism and its application to Singapore. *Environment and Urbanization ASIA*, 1(2), 149–170. https://doi.org/10.1177/097542531000100204

Ng, M. K. (2019). Governing green urbanism: The case of Shenzhen, China. *Journal of Urban Affairs*, 41(1), 64–82. https://doi.org/10.1080/07352166.2016.1271623

Okello, J. A., & Akoko, J. A. (2023). Smart city trends and innovations shaping the future of cities. *Scientific Papers of Silesian University of Technology. Organization and Management Series*, *2023*(169), 553–561. https://doi.org/10.29119/1641-3466.2023.169.33

Patacchini, E., & Zenou, Y. (2009). Urban Sprawl in Europe. Brookings-Wharton Papers on Urban Affairs 2009, 125–149. https://dx.doi.org/10.1353/urb.2009.a271142

Rafael, S., Martins, H., Sá, E., Carvalho, D., Borrego, C., & Lopes, M. (2016). Influence of urban resilience measures in the magnitude and behaviour of energy fluxes in the city of Porto (Portugal) under a climate change scenario. *Science of the Total Environment*, *566–567*, 1500–1510. https://doi.org/10.1016/j.scitotenv.2016.06.037

Raghubanshi, A. S. (2024). Sustainable urban planning: A comparative study of green city initiatives around the worldqq. *Journal of Sustainable Solutions*, 1(3), 18–23. https://doi.org/10.36676/j.sust.sol.v1.i3.16

Rai, P., & Rai, S. S. (2024). The vedic texts on sustainable development in ancient south asia. *GeoJournal*, 89(3), 90. https://doi.org/10.1007/s10708-024-11092-9

Raikes, J., Smith, T. F., Baldwin, C., & Henstra, D. (2022). The influence of international agreements on disaster risk reduction. *International Journal of Disaster Risk Reduction*, 76, 102999. https://doi.org/10.1016/j.ijdrr.2022.102999

Ramesh, M. N., & Nijagunappa, R. (2015). Green roofs – an eco-friendly approach to sustainable livelihood. *Environmental Science: An Indian Journal*, 1(1), 1–. https://www.tsijournals.com/abstract/green-roofs-an-ecofriendly-approach-to-sustainable-livelihood-3102.html

Redpoll Way and Redstart Way in Abbeymead, Gloucester. (2015, August 28). Wikimedia Commons. https://commons.wikimedia.org/wiki/File:Redpoll_Way_and_Redstart_Way_in_Abbeymead,_Gloucester.JPG

Rezvani, S., de Almeida, N., & Falcão, M. (2023). Climate adaptation measures for enhancing urban resilience. *Buildings*, 13(9), 2163. https://doi.org/10.3390/buildings13092163

Romanelli, M. (2021). Driving smart cities through smart projects. *Management Trends in the Context of Industry 4.0*. https://doi.org/10.4108/eai.4-12-2020.2303264

Romero-Muñoz, S., Sánchez-Chaparro, T., Muñoz Sanz, V., & Tillie, N. (2024). Urban greening management arrangements between municipalities and citizens for effective climate adaptation pathways: Four case studies from The Netherlands. *Land*, *13*(9), 1414. https://doi.org/10.3390/land13091414

Rozhkov, A. (2024). *Decentralized renewable energy integration in the urban energy markets: A system dynamics approach*. https://doi.org/10.21203/rs.3.rs-3835543/v1

Ruiz-Vanoye, J. A., Díaz-Parra, O., Ruiz-Jaimes, M. A., Ortiz, J. A., Simancas-Acevedo, E., Rodriguez-Flores, J., Martínez Mireles, J. R., Xicoténcatl-Pérez, J. M., & Vera Jimenez, M. A. (2024). The role of smart cities in creating sustainable urban environments through strategies and innovative solutions. In G. Gülhan & O. Polat (Eds.), *Interdisciplinary approaches to transportation and urban planning* (pp. 23–46). IGI Global Scientific Publishing. https://doi.org/10.4018/979-8-3693-6695-0.ch002

Sahni, S., & Aulakh, R. S. (2014). Planning for low carbon cities in India. *Environment and Urbanization Asia*, 5(1), 17–34. https://www.researchgate.net/publication/270709270_Planning_for_Low_Carbon_Cities_in_India

Sahoo, B. K. (2020). Reinventing Gandhian ideas on sustainable development in the age of environmental crisis. *Asiatic Society for Social Science Research (ASSSR)*, 2(1), 15. DOI:10.46700/asssr/2020/v2/i1/196109

Sande, & Phadtare. (2015). Comparative study of LEED and GRIHA rating system. *Journal of Information, Knowledge and Research*, *3*(2), 168–174.

Sareen, S., & Waagsaether, K. L. (2023). New municipalism and the governance of urban transitions to sustainability. *Urban Studies*, 60(11), 2271–2289. https://doi.org/10.1177/00420980221114968

Schuetze, T., & Chelleri, L. (2013). Integrating decentralized rainwater management in urban planning and design: Flood resilient and sustainable water management using the example of coastal cities in the Netherlands and Taiwan. *Water*, 5(2), 593–616. https://doi.org/10.3390/w5020593

Schumacher, E. F. (1973). *Small is beautiful: A study of economics as if people mattered*. Blond & Briggs. https://doi.org/10.2307/2615239

Shahrabani, M. M. N., & Apanavičienė, R. (2022). *Towards integration of smart and resilient city: literature review*. IOP Conference Series: Earth and Environmental Science, *1122*(1), 012019. https://doi.org/10.1088/1755-1315/1122/1/012019

Sharma, K., A K, D., Upadhyay, R., & Kulhar, K. S. (2024). *Review of renewable energy, sustainability concerns, and climate solutions for smart cities*. E3S Web of Conferences, 540, 13004. https://doi.org/10.1051/e3sconf/202454013004

Sijakovic, M., & Peric, A. (2021). Sustainable architectural design: towards climate change mitigation. *Archnet-IJAR: International Journal of Architectural Research*, 15(2), 385–400. https://doi.org/10.1108/ARCH-05-2020-0097

Smith, R. M., & Pathak, P. (2018). Urban sustainability in India: Green buildings, AMRUT Yojana, and smart cities. In B. Grant, Liu, & L.s.), *Meropolitaovernance in Asia and the Pacific Rim* (pp. 163–190). Springer Singapore. https://doi.org/10.1007/978-981-13-0206-0_9

Social equity for sustainable cities | U.S. Green Building Council. (2025). USGBC. https://www.usgbc.org/articles/social-equity-sustainable-cities

Stagno, E. (2021). Investigating relationship between built environment and health. In C. Bevilacqua, F. Calabrò, & L. Della Spina (Eds.), *New metropolitan perspectives. NMP 2020. Smart innovation, systems and technologies* (vol. 178, pp. 285–296). Springer. https://doi.org/10.1007/978-3-030-48279-4_27

Swain, A., Salkuti, S. R., & Swain, K. (2021). An optimized and decentralized energy provision system for smart cities. *Energies*, *14*(5), 1451. https://doi.org/10.3390/en14051451

Taneja, S., Jaggi, P., Jewandah, S., & Ozen, E. (2022). Role of social inclusion in sustainable urban developments: An analyse by PRISMA technique. *International Journal of Design & Nature and Ecodynamics*, 17(6), 937–942. https://doi.org/10.18280/ijdne.170615

The Historic England Blog. (2023, December 8). *A brief introduction to garden cities*. https://heritagecalling.com/2016/02/18/a-brief-introduction-to-garden-cities/#:~:text=Howard's%20vision,sufficient%20as%20far%20as%20possible.

Thoreau, H. D. (1854). *Walden*. Ticknor and Fields.Turner, B., Devisscher, T., Chabaneix, N., Woroniecki, S., Messier, C., & Seddon, N. (2022). The role of nature-based solutions in supporting social-ecological resilience for climate change adaptation. *Annual Review of Environment and Resources*, 47(1), 123–148. https://doi.org/10.1146/annurev-environ-012220-010017

United Nations. (2018). *2018 Revision of world urbanization prospects*. Department of Economic and Social Affairs. https://www.un.org/development/desa/en/news/population/2018-revision-of-world-urbanization-prospects.html

United Nations. (2022). *World Population Prospects 2022: Summary of Results*. https://www.un.org/development/desa/pd/sites/www.un.org.development.desa.pd/files/wpp2022_summary_of_results.pdf

Urban Design lab. (2023, July 28). *Book review: The garden cities of tomorrow by Ebenezer Howard*. Urban Design lab. https://urbandesignlab.in/book-review-the-garden-cities-of-tomorrow-by-ebenezer-howard/

van den Bosch, M., & Ode Sang, Å. (2017). Urban natural environments as nature-based solutions for improved public health – A systematic review of reviews. *Environmental Research*, 158, 373–384. https://doi.org/10.1016/j.envres.2017.05.040

Verdeil, E. (2019). Arab sustainable urbanism: Worlding strategies, local struggles. *Middle East – Topics & Arguments, Urban Development, 12*(12), 35–42. https://doi.org/10.17192/meta.2019.12.7935

Vezzoli, C., Ceschin, F., Osanjo, L., M'Rithaa, M. K., Moalosi, R., Nakazibwe, V., & Diehl, J. C. (2018). Distributed/decentralised renewable energy systems, 23–39. https://doi.org/10.1007/978-3-319-70223-0_2

Wiratama, Y. H., Sobari, W., & Mashuri, A. (2023). Penerapan Tata Kelola Lingkungan dalam Pengelolaan Ruang Terbuka Hijau (RTH) di Kabupaten Ponorogo. *Brawijaya Journal of Social Science*, 2(02). https://doi.org/10.21776/ub.bjss.2023.002.02.3

Zhong, J., Shao, X., Xiao, H., Yang, R., & An, X. (2023). The research on the green leadership: A systematic review and theoretical framework. *Environment, Development and Sustainability*, *27*(1), 377–08. https://doi.org/10.1007/s10668-023-03960-0

Zuberi, F. (2017). Green urbanism in medieval India, Agra Fort: A sustainable eco-city. *Procedia Environmental Sciences*, 37, 12–21. https://doi.org/10.1016/j.proenv.2017.03.009

2 Striving for Carbon Neutrality

Advancing from Nature-Based to Nature-Driven Urbanism

Nazuk Bhasin, Sudhanshu Kumar, Amit Kumar Tiwari, Anil Barla, and Gopal Shankar Singh

2.1 THE CONCEPT OF URBANISM

Urbanism, the study of interactions between city dwellers and their built environments, integrates urban planning, architecture, sociology, and geography to enhance residents' quality of life. With the evolution of civilization, from prehistoric societies to contemporary metropolises, urbanism has witnessed a significant transformation. In the early stages of urbanism, infrastructure and sanitation were prioritized, as evidenced by the well-planned settlements of the Indus Valley and ancient Greek civilizations, thus, establishing the foundations for structured urban development (Spreiregen, 1965). However, it was the Industrial Revolution which prompted a transformation in the course of history. Megacities like London, characterized by overpopulation and poor infrastructure, emerged as a consequence of swift urbanization caused by the disruptive forces of Industrial Revolution (Hobsbawm, 1999). Urban planning emerged as a discipline in response, with concepts like Ebenezer Howard's "Garden City" pioneering sustainable urban development (Hall, 2014). However, modernist planning, epitomized by Le Corbusier's "Radiant City", faced criticism for neglecting social considerations (Le Corbusier, 1967). Jane Jacobs advocated for mixed-use development and community involvement, emphasizing social cohesion in urban planning (Jacobs, 1961). Contemporary urbanism focuses on sustainability and smart cities, with initiatives like green building practices and digital infrastructure driving urban development (WGBC, 2018). Smart city projects, exemplified by Barcelona, Singapore, and Amsterdam, leverage technology to enhance urban efficiency and quality of life (Hollands, 2008). The relevance of social justice continues to rise as evidenced by programs like Portland's "10-Minute Neighbourhood" and Brazil's "Right to the City" movement, which seek to build inclusive communities

DOI: 10.1201/9781003518037-3

(Friendly, 2013; Miles & Song, 2009). In cities such as Chicago and Porto Alegre, participatory budgeting gives citizens the ability to influence urban policy, encouraging neighbourhood involvement and fair development (de Sousa Santos, 1998). Urbanism continues to evolve, driven by technological advancements, environmental concerns, and social aspirations, shaping the cities of tomorrow.

2.1.1 Aspects of Urbanism

Urbanism, as a comprehensive field of study, encompasses a myriad of aspects that collectively define the character, functionality, and vitality of urban environments. This multifaceted field addresses not only the physical infrastructure and layout of cities but also the social, economic, and cultural dynamics that animate urban environments. The fundamental elements that contribute to the vitality, sustainability, and inclusivity of modern urban spaces need to be understood to gain further insights into the process of urbanism. Urban planning, integral to urbanism, involves organizing urban spaces through policies on land use, transportation, and amenities. With urban populations expected to surge, effective planning is crucial. According to recent estimates by the World Bank, urban areas are projected to absorb an additional 2.5 billion people by 2050 (Huang et al., 2019). Cities globally showcase innovative urban planning. For example, Seoul revitalized its Cheonggyecheon stream, enhancing public space. Medellín improved access to marginalized areas through cable cars and escalators, fostering social inclusion. Vancouver prioritized sustainability through green spaces and transit-oriented development. These initiatives demonstrate diverse approaches to urban planning for creating liveable and resilient cities (Figure 2.1).

Architecture is pivotal in urbanism as it moulds the tangible appearance of the city, encompassing the creation of edifices and constructions that enrich the

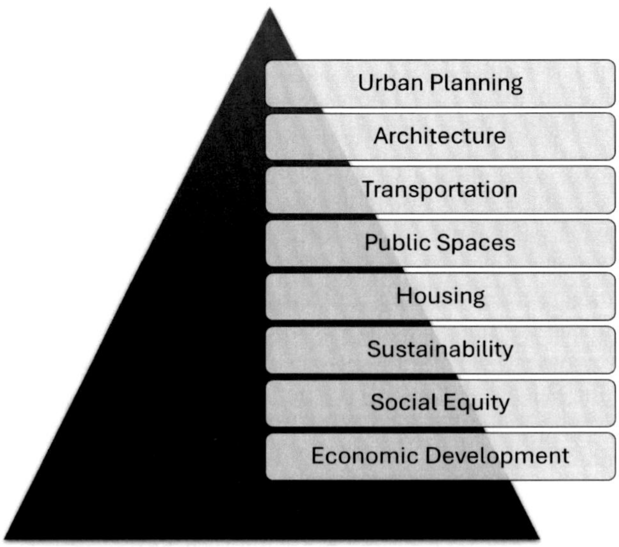

FIGURE 2.1 Aspects of urbanism.

urban scenery and cater to the populace's requirements. The operationality of urban regions heavily relies on effective transportation systems, prompting urbanism to explore diverse transportation methods like public transit, cycling, pedestrianism, and private vehicles to devise cohesive and sustainable mobility alternatives. The utilization of public transit can markedly diminish greenhouse gas emissions, with metropolitan centres such as New York and Tokyo reaping the benefits of extensive and efficient transit infrastructures (Stanley et al., 2023).

Urbanism focuses on creating accessible, aesthetically pleasing, and valuable public spaces. Public spaces, such as parks, plazas, and streetscapes, give residents places for leisure, social contact, and community events. Well-designed public spaces can improve social cohesiveness and reduce crime rates (Crabill, 2009). A key aspect of urbanism is ensuring a supply of housing that is diversified, affordable, and of high quality, taking into account factors like density, zoning, and housing policy. According to the World Resources Institute, the proportion of urban residents without access to affordable housing is predicted to increase by more than 30% to 440 million households, or 1.6 billion people by 2025 (King et al., 2017).

Through the promotion of green infrastructure, compact development, the integration of renewable energy, effective waste management, climate resilience, and the protection of natural resources, urbanism also seeks to create environmentally sustainable cities. Over 60% of the world's energy is consumed in cities, and they are accountable for almost 70% of the world's greenhouse gas emissions, according to the International Energy Agency's 2022 study. Thus, urbanism is vital for minimizing carbon footprints and mitigating the effects of climate change. Additionally, urbanism strives towards creating inclusive cities that address social injustices, provide access to fundamental services, and promote community involvement, offering every citizen equal opportunity. Cities with higher levels of social fairness typically experience swifter economic growth and higher standards of living (Angelopulo, 2021).

Urbanism also aims to foster economic vitality by assisting small-scale businesses, drawing in investment, and generating employment, in order to create resilient economies that can adjust to changing circumstances. The economic significance of well-planned urban areas is highlighted by the fact that metropolitan areas account for an impressive 84% of the global Gross Domestic Product (Chatterjee et al., 2020; King, 2015).

The analysis of web of science (WoS) records of published articles on "urbanism" reveals that 40.92% of the publications addressed SDG 11: Sustainable Cities and Communities. Publications on the topic are mostly classified under WoS research areas "urban studies", "geography", "environmental studies" and "architecture". This research topic has been mostly explored and published by researchers and scientists from United States of America, England, and Canada (Clarivate, 2022).

The analysis of term co-occurrence based on text data of articles on urbanism revealed co-occurrence of "urbanism", "city", "architecture", "urban planning", "gentrification", "sustainable urbanism", "landscape", "planning", "ecosystem service" and many more (Figure 2.2) in titles and abstract of the articles (Van Eck & Waltman, 2010) which shows thematic relationship with the words occurring together (Donthu et al., 2021).

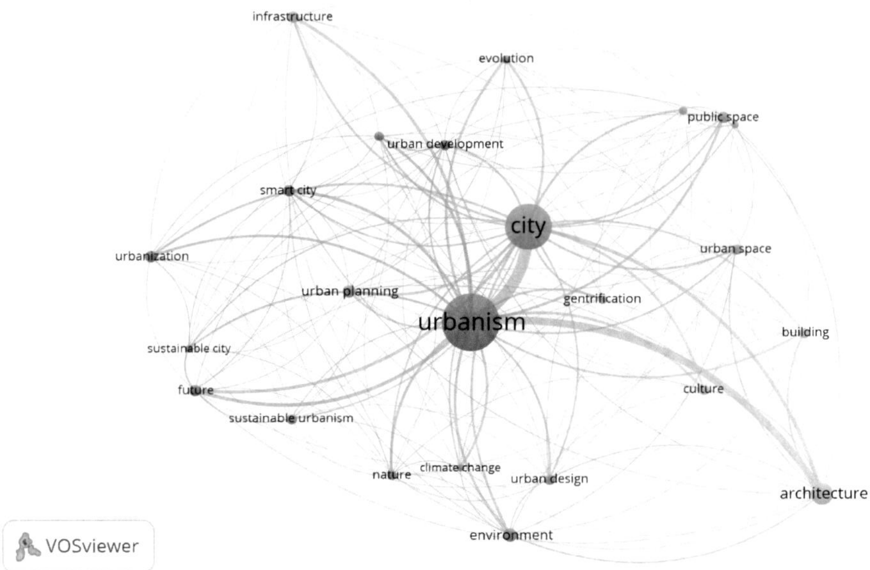

FIGURE 2.2 Text co-occurrence within titles and abstracts of articles on urbanism.

The challenges posed by hasty and precipitous urbanization, including overcrowding, contamination, scarce infrastructure, and inefficient resource use, can be effectively tackled by studying the concept of urbanism. The application of these concepts to urban planning and management can render urban centres to become sustainable, robust and habitable while focusing on furthering the complete welfare of the residents.

2.1.2 Distinguishing Urbanization from Urbanism

Urbanization and urbanism are intricately intertwined theories as both deal with the dynamism and advancement of cities. However, they take distinctive approaches to different facets of urban growth and living. Urbanization can be defined as the demographic and tangible progression of cities. The drivers of urbanization include economic considerations, access to resources and amenities and improved living paradigms, consequently causing alterations in the land use patterns, economic and societal arrangements. For instance, Shenzhen, despite facing issues like inadequate housing and income disparity, has evolved from being a fishing settlement to a major global technology and innovation hub, while witnessing a population growth from 30,000 to over 12 million since 1980 (Shenzhen Statistics Bureau, 2021).

Contrarily, urbanism is fixated on accentuating the qualitative attributes of urban existence. It is primarily concerned with facets such as planning, architecture and sustainability, which are aimed at designing proficient and habitable urban environments. Several smart cities such as Amsterdam, Barcelona, Singapore have leveraged technology in order to enhance the quality of life for their citizens, by upgrading the infrastructure, resources, and services delivery (Caragliu et al., 2011). However, New

Urbanism, emerging in the late 20th century, counters urban sprawl and promotes walkable, mixed-use, and transit-oriented development. Examples include Seaside and Celebration, Florida, which exemplify these principles (Duany et al., 2000; Calthorpe & Fulton, 2001).

The interaction and association between urbanization and urbanism manifests itself in the necessity to effectively design urban spaces so as to accommodate the population growth and the associated socio-economic and environmental challenges. Urbanism approaches have been utilized by cities like Copenhagen and Amsterdam to encourage sustainable commutation. Efficacious planning strategies in urban areas can help reduce the energy intake and the emanations, thus, attending to the environmental challenges. Cities are responsible for consuming 78% of the global energy and 75% of the global greenhouse gas emissions (Verma et al., 2021). Urbanism not only concentrates on the economic and environmental viability but also promotes social equity by incorporating projects like community land trusts, affordable housing schemes, thus, furthering the idea of inclusive cities (Fink, 2019). In India, Pradhan Mantri Awas Yojana, which works on the principle of providing affordable housing to all residents by 2024, is a good example to illustrate the social equity paradigm of urbanism (Sanga, 2024).

Hence, urbanism is a broader notion incorporating socio-economic, cultural and environmental dimensions of urban existence whereas urbanization only pertains to the physical and demographic progression of urban areas. Urbanism further extends to the planning, optimization and management required to develop cities as inclusive, sustainable and efficient centres. Both these concepts together contribute to the qualitative enhancement and quantitative expansion of urban areas and are indispensable for the prospects of urban living. Comprehending the divergences between urbanization and urbanism is imperative in order to confront the multidimensional impediments and prospects linked to urban development in the 21st century. These include the need to manage swift urban expansion and guarantee unbiased access to resources in order to establish sustainable, liveable cities for all inhabitants.

2.1.3 Deciphering the Urbanism Quandary: Exploring its Impact

Urbanization and urbanism exert profound influence across myriad dimensions, spanning societal, economic, cultural, and environmental spheres. In terms of socio-economic impact, urbanization acts as a catalyst for population growth, concentrating individuals in cities and driving significant demographic shifts (United Nations, 2018). Economically, cities emerge as vibrant centres of commerce, innovation, and opportunity, attracting substantial investments, businesses, and skilled labour (Florida, 2002). This growth stimulates job creation across diverse sectors, leading to higher incomes and improved living standards for urban residents (Glaeser, 2011). Cities such as New York and Tokyo are primary examples of economic motivation, with their prospering financial districts, sprightly cultural landscapes, and locations with cutting-edge technology innovation (Florida, 2002). In addition to drawing talent and investments, the vivacious economies of these cities also nurture entrepreneurship and worldwide economic expansion.

Additionally, urban areas recurrently provide opportunities for poverty alleviation through improved access to economic prospects, healthcare, and educational services, and act as hubs for social mobility (Chetty et al., 2014). People residing in urban settlements are more likely to witness upward social mobility as compared to residents of rural areas (Chetty et al., 2014). Yet, there are concerns with infrastructure and service delivery brought on by the rapid population growth of cities. Strong transit systems, effective utilities and services, and sufficient housing solutions are essential to aid the rising population (Glaeser, 2011). For example, to tackle population increase while also ensuring a high standard of living for citizens, cities like Singapore and Copenhagen have made substantial investments in sustainable urban design and public transportation systems (Glaeser, 2011).

Urbanization has far-reaching and multifaceted impacts on environment altering ecosystems and landscapes and also influencing the resource dynamics. Expansion of urban areas frequently ensues in the encroachment of agricultural and natural ecosystems, laying the foundation for habitat fragmentation and loss, and a reduction in biodiversity (Seto et al., 2012; McDonald et al., 2008). This transformation of land use disrupts ecological processes and ecosystem services like pollination, flood control, and water purification and detrimentally impacts the surrounding flora and fauna (Foley et al., 2005). Moreover, the rapid expansion of urban areas places tremendous strain on natural resources such as raw materials, energy, and water (Grimm et al., 2008). Unabated resource consumption aggravates problems, including soil erosion, deforestation, and water scarcity instigated by resource depletion and environmental degradation (Angel et al., 2011; Seto et al., 2012). Furthermore, the intensity of anthropogenic activity in cities causes emissions of pollutants such as heat, particulate matter, and greenhouse gases (Singh et al., 2020; Kennedy et al., 2015). These pollutants contribute to exacerbating global environmental issues like climate change and urban heat islands while supplementing the deterioration of the quality of the environment (Oke, 1982).

Urbanization presents both obstacles as well as opportunity on the socio-cultural front. Cities entice a varied population and foster thriving urban communities by functioning as nuclei for ingenuity, innovation, and cross-cultural interchange (Mang, 2019). However, social inequality may also face deterioration by the quickening speed of urban development, resulting in physical segregation, marginalization, and exclusion (Jones, 2023; Lees, 2008). Urban redevelopment initiatives frequently result in gentrification and the displacement of low-income neighbourhoods, thus, threatening social cohesiveness and community resilience (Atkinson & Bridge, 2005).

Sustainable urban design and governance are significant aspects in confronting these intricate concerns. Urban resilience can be enhanced and environmental effects can be abridged with the application of techniques like compact urban development, green infrastructure, and sustainable mobility (UN, 2016). Furthermore, encouraging inclusive and participatory methods for urban development can prove beneficial to ensure that the advantages of urbanization are fairly distributed among the population, empowering underprivileged and vulnerable communities, and promoting social fairness (Hollands, 2008).

Therefore, urbanization and urbanism are transformational forces that influence economies and civilizations globally. They propel population expansion, economic growth, and social advancement. Conversely, they are responsible for infrastructure and service delivery issues. Thus, to maximize the benefits of urbanization while minimizing its adverse impacts, sustainable infrastructure investment and efficient urban planning in the form of urbanism are essential.

2.2 NECESSITY FOR CARBON NEUTRALITY: THE URBAN IMPERATIVE

Realizing carbon neutrality within the frameworks of urbanization and urbanism has become indispensable for mitigating global warming, protecting public health, building resilience, and maximizing economic potential. Greenhouse gas emanations have continued to intensify globally, with 2023 establishing a new high record (Forster et al., 2024). Urban areas persist to be considerable contributors to the greenhouse gas discharges despite extensive advances, thus, emphasizing the unrelenting necessity for sustainable urban solutions (UNEP, 2023). To assuage the deleterious consequences of global warming, like extreme weather episodes and ascending sea levels, a transfer to carbon-neutral urban areas is required (IPCC, 2018). Furthermore, air pollution in cities is frequently associated with release of greenhouse gases in the atmosphere, causing an increased possibility of respiratory disorders and premature mortality among the residents (Manisalidis et al., 2020). Cities may work towards enhancing human health and improving air quality by effecting sustainable commutation alternatives and adoption of renewable energy sources (WHO, 2019). Urbanism driven by carbon neutrality aspires to transform cities into resilient centres by employing policies such as climate-adaptive architecture and green infrastructure. To exemplify the feasibility of accomplishing carbon neutrality, cities such as Copenhagen, Denmark, have fixed aspiring objectives for themselves by apportioning resources for sustainable mobility infrastructure and renewable energy (Sovacool et al., 2020). Apart from reducing climate risks, cities prioritizing carbon neutrality entice investment and are primary promoters for the progression of the principles of a green and circular economy (IRENA, 2021). The implementation of carbon neutrality for the development of urban areas helps address the challenges posed by climate change, promotes sustainability and ensures a resilient and affluent future for urban inhabitants.

2.3 NATURE-BASED URBANISM: DRIVERS OF CARBON NEUTRALITY

Nature-based urbanism (NBU) has garnered importance for attaining carbon neutrality within urban ecosystems, because it integrates natural elements and processes to transform urban design and planning (Romanovych, 2022). NBU leverages the natural capability of the ecosystems to mitigate carbon emanations and enhance carbon sequestration (Bayulken et al., 2021). It includes sustainable landscaping, urban forestry, and green infrastructure in its scaffold (Snep et al., 2023). The combination of these strategies are employed to reduce the carbon intensity of urban landscapes while fortifying their resilience to climate change (Figure 2.3).

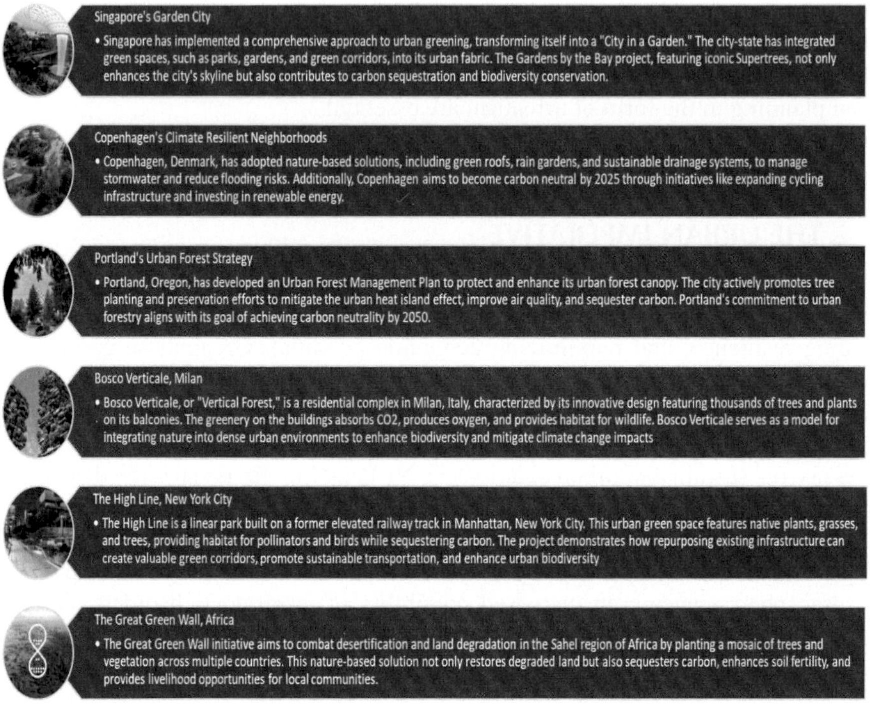

FIGURE 2.3 International best practices incorporating nature-based solutions for urbanism.

Green infrastructure is crucial in NBU, featuring elements like green roofs, permeable pavements, and rain gardens. These not only manage stormwater runoff and reduce urban heat islands but also absorb CO_2 (Jones et al., 2022; Song, 2022). Green roofs significantly offset emissions from buildings and transportation (Nowak et al., 2014).

Urban forests and green spaces also play a vital role in carbon sequestration. Trees absorb CO_2 through photosynthesis, storing it in biomass and soil. Additionally, urban forests improve air quality by filtering pollutants, with trees removing airborne particulates, nitrogen dioxide, sulphur dioxide, and carbon monoxide. In Beijing, urban trees reduced suspended particulate matter by 9%, enhancing air quality and reducing respiratory issues (Gao et al., 2020). Green spaces mitigate the urban heat island effect. Trees and vegetation provide shade and cool air through transpiration. In Singapore, increasing urban greenery reduced ambient temperatures by 4°C (Wong et al., 2021). Additionally, green spaces support biodiversity (Aerts et al., 2018).

Access to urban green spaces enhances mental and physical well-being. Exposure to nature reduces stress, anxiety and depression (Bratman et al., 2019). Physical activities like walking, jogging, and cycling in green spaces improve health, reducing chronic disease risks. Bogotá's "Ciclovía" closes 120 km of city streets to motor vehicles weekly, promoting physical activity, reducing air pollution, and fostering community interaction (Mejia-Arbelaez et al., 2021). Endorsing activities like walking and bicycling as modes of active transportation supplementing public transportation lowers carbon emissions and reliance on fossil fuel-operated automobiles. For instance, in the case of European cities, substituting automobiles by bicycles could minimize CO_2 emissions by nearly about 10% (Gössling et al., 2019). A meta-analysis by Saunders et al. (2013) found that individuals who are physically active and commute actively are healthier than those who commute by car. Improvement in carbon sequestration capacity, encouragement of physical exercise, and lessening of traffic congestion can be arrived at simultaneously by constructing bike lanes and pedestrian routes bounded by vegetation (Ferrini et al., 2020). The adoption of active transportation modes and a robust public transport network contribute to alleviation of traffic congestion further ensuring an efficient and effortless traffic flow and less idling, thus lowering air pollution (Cheng et al., 2020). For example, construction of bike lanes and proliferation of public transportation in Los Angeles have assisted in reducing traffic and enhancing air quality (T.D.M, 2021).

Nature-based solutions, such as wetland restoration, riparian buffer zones, and reforestation, enhance natural habitats and boost ecosystem services vital for urban sustainability. Coastal wetlands, including mangroves and salt marshes, are particularly effective carbon sinks, sequestering carbon at rates up to four times higher than terrestrial forests (Mitsch & Gosselink, 2015). For example, the South Bay Salt Pond Restoration Project in the San Francisco Bay Area is aimed at restoring tidal marsh habitats, augmenting carbon sequestration, elevating water quality, and conserving endangered species (Adam, 2019). Wetland restoration aims at benefitting communities by promoting biodiversity, reducing the influence of climate change, and extending spaces for recreation and education (Alikhani et al., 2021).

Another approach for enhancing urban sustainability is urban agriculture. It includes rooftop gardens, urban farms and community-supported agriculture (CSA) programs, which offer comprehensive solutions to minimize the carbon footprint of the food production and supply chain (Ackerman et al., 2014). In addition to positively impacting the carbon emissions, it further contributes to improving the food security. This is achieved by focusing on food production closer to its consumers, thus, eliminating the logistics cost incurred while transportation of food items. Consequently, this reduces the fuel and energy bill. For instance, urban agriculture has the potential to cut greenhouse gas emissions accompanying the food transport by approximately 11% (Kulak et al., 2013). Moreover, urban agriculture approaches transmute otherwise vacant areas into fruitful gardens, furthering the objective of efficient and innovative land use in urban areas. For example, the "Urban Leaves" project operational in Mumbai, India, supplies fresh produce to urban residents and contribute effectively to sustainable and vigorous food schemes (Akaeze & Nandwani, 2020). The benefits of rooftop gardens can further be extended to providing insulation to buildings, minimizing the energy utilisation and controlling rainwater (Amos et al., 2018). Rooftop gardens were found to reduce indoor temperatures by nearly 3–4°C in Singapore (Wong et al., 2003). Urban farms also generate employment and help create awareness and promote educational initiatives among locals (Allen & Clawson, 2018). In Detroit, Michigan, initiatives like the "D-Town Farm" contribute to local food systems while providing employment opportunities and educational resources for the community (Taylor & Ard, 2015).

Another remarkable instance of the application of NBU to urban ecosystems is the Sustainable Urban Design. It employs environmental-friendly and green resources; technology driven by energy-efficiency and passive design policies and is aimed at reducing the energy utilization (Nižetić et al., 2019) and the carbon footprint of urban areas. A 90% reduction in carbon emanations can be achieved when recycled steel and reclaimed wood are used as construction materials in place of conventional materials (Chastas et al., 2016). Including smart building management systems, light-emitting diodes (LEDs) and other energy-efficient solutions can lower the energy consumption by 30–50% (Pode, 2020). The Bedding Zero Energy Development (BedZED) in London is an outstanding exemplification of sustainable urban design. It incorporates elements like solar panels and stormwater harvesting systems to achieve carbon neutrality (BioRegional, 2020). Hammarby Sjöstad in Stockholm and the CII-Sohrabji Godrej Green Business Centre in Hyderabad are other notable examples. They integrate sustainable materials, renewable energy sources, and efficient waste management systems, achieving substantial reductions in energy consumption and water use (Singh et al., 2023; Pandis Iveroth et al., 2013).

The decline in carbon footprint, waste generation capacity and resource utilisation efficiency of urban populations can be achieved by embracing the principles of circular economy (Velenturf & Purnell, 2021). The basic tenets governing these principles are the practices of reduce, reuse, recover, regenerate and recycle (Ellen MacArthur Foundation, 2020). These approaches are in turn aligned with the essence of NBU (Kabisch et al., 2022). Waste reduction and management policies and schemes like recycling and composting, are imperative for reducing greenhouse gas emissions, specifically the emissions of methane from landfill locations (Brown et al., 2008). The adverse environmental impacts of urban areas can be minimized by inculcating innovative methods that impersonate natural ecosystems promoting the circularity paradigm while simultaneously augmenting the ecosystem services, biodiversity and sustainability (McPhearson et al., 2016). Sweden's programme on achieving an absolute recycling rate (Avfall Sverige, 2021) and Japan's "3R" initiative (Japan Ministry of the Environment, 2020) are some of the efficacious strategies construed to further the goal of sustainable development in the context of urban environments (EPA, 2020).

Apart from the environmental benefits offered by NBU, it also promotes social cohesion (Jennings & Bamkole, 2019). Urban green spaces function as centres for the promotion of collaboration, communication and recreation, fostering a sense of fraternity and community (Klinenberg, 2018). Neighbourhood social contacts were found to be enhanced when citizens participated in community gardening initiatives in the city of Chicago (Okvat & Zautra, 2011). Moreover, CSA programmes were found to be establishing an undeviating direct association between the producers and the consumers, thus, promoting community resilience (CoDyre et al., 2015). Awareness generation and engagement programs centring nature-based solutions for carbon neutrality, at the community level, can prove helpful to achieve the objectives of sustainable urban development by promoting environmentally responsible procedures (Mahmoud et al., 2023).

The analysis of WoS records of published articles on "nature-based urbanism" reveals 46.95% of the publications addressed SDG 11: Sustainable Cities and

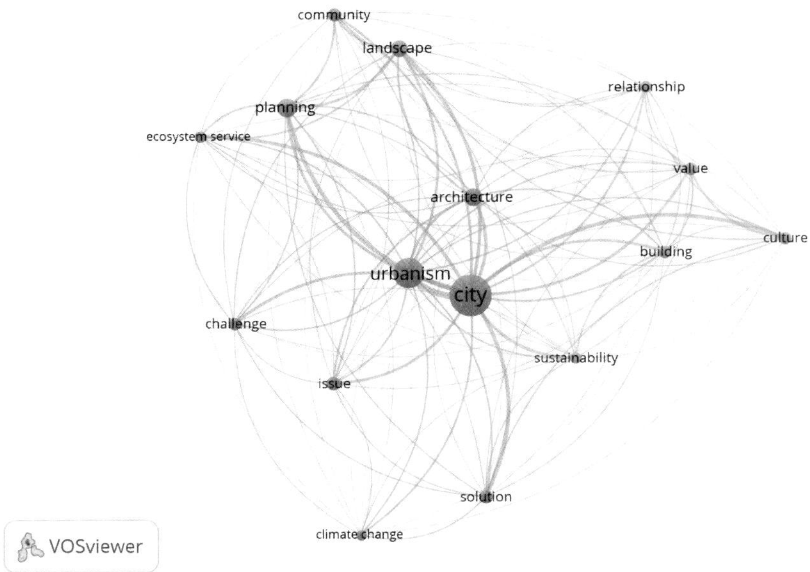

FIGURE 2.4 Text co-occurrence within titles and abstracts of articles on NBU.

Communities, remaining addressed SDG 13: Climate Action and SDG 15: Life on Land. Publications on the topic are mostly classified under WoS research areas "Environmental Studies", "Environmental Sciences", and "Urban Studies". This research topic has been mostly explored and published by researchers and scientists from England and China (Clarivate, 2022)

The analysis of term co-occurrence map based on text data of titles and abstract of the articles on NBU revealed co-occurrence of "urbanism", "city", "architecture", "sustainability", "landscape", "planning", "ecosystem service", and many more (Figure 2.4) (Van Eck & Waltman, 2010) highlighting the relationship between the concept of NBU and the words co-occurring in published literature (Donthu et al., 2021).

Quintessentially, NBU offers a comprehensive, holistic method to transform urban development towards a more carbon-neutral developmental paradigm. By integrating and incorporating the natural capital and capabilities, within the urban ecosystems, goals of climate change resilience and reduction in the carbon footprint of cities can be attained while fostering robust, dynamic and sustainable environments.

2.4 WAY FORWARD: NATURE-DRIVEN URBANISM

Nature-driven urbanism (NDU) stands as a refined iteration of NBU strategies. NDU considers cities as complex ecosystems interconnected with surrounding landscapes, addressing environmental challenges like carbon emissions, biodiversity loss, and climate resilience simultaneously (McPhearson et al., 2016). The NDU approach aims at active contribution to ecological restoration and regeneration while incorporating the reformative design principles encouraged by natural ecosystems and working at

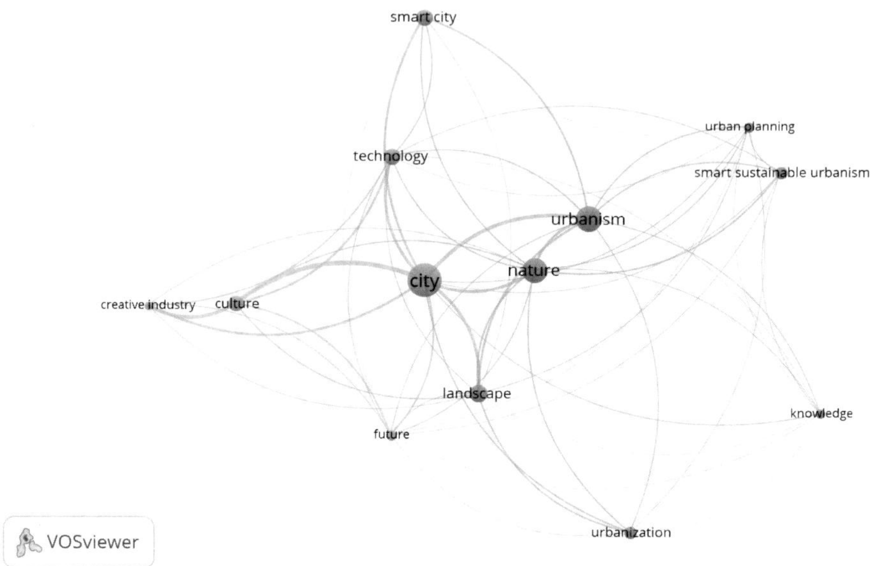

FIGURE 2.5 Text co-occurrence within titles and abstracts of articles on nature-driven urbanism.

enhancing the ecosystem functionalities and services (Roggema, 2020). It prioritizes creation of a network by associating urban ecosystems with adjoining environments for the formation of ecological corridors aimed at supporting biodiversity, thereby facilitating its movement (Roggema, 2021). Further, it enhances the resilience of the ecosystem by promoting ecosystem services such as air quality improvement, climate regulation and carbon sequestration (Roggema, 2023). NDU also contributes to further the principles of circularity in context of the urban environment so as to improve the efficacy of resource consumption, reduce the waste production and associated carbon releases (Langergraber et al., 2020). Cities can assist a united front towards achieving the objectives of carbon neutrality and sustainability by encouraging social cohesion, mental and general well-being among inhabitants by inclusive and informed decision-making practices and inculcating a sense of oneness with nature (Kabisch et al., 2022). In essence, NDU offers a transformative vision for creating cities that are not only carbon-neutral but also resilient, regenerative, and in harmony with the natural world, ensuring a sustainable future for generations to come.

The analysis of WoS records of published articles on "nature-driven urbanism" revealed that 50% of the publications on NDU addressed SDG 11: Sustainable Cities and Communities, whereas a smaller number of articles dealt with SDG 13: Climate Action and SDG 15: Life on Land. Majority of the publications are classified under the WoS research areas "Environmental Sciences Ecology" and "Urban Studies", whereas a large contribution to the topic is from the authors and researchers based in England and Australia (Clarivate, 2022) (Figure 2.5).

The analysis of term co-occurrence map based on text data of titles and abstract of the articles on NDU revealed co-occurrence of "urbanism", "nature", "city", "smart city", "technology", "smart sustainable urbanism", "urbanization", "urban planning", and others (Van Eck & Waltman, 2010) signifies relationship and importance of various concepts involved.

Fundamentally, both NDU and NBU emphasize on the integration of natural elements into urban ecosystems as a means of fostering resilience and sustainability, yet the inherent divergence lies in their ideology and approach. NBU characteristically incorporates natural components and practices with urban environments to promote and strengthen the ecosystem services like rainwater management, microclimate regulation (McPhearson et al., 2016). Urban plantations, green roofs, and other ideas that advance urban ecosystems and prove advantageous to urban residents illustrate the functionalities and the vision of NBU. NBU considers nature as a means to accomplish particular objectives with urban planning, optimization and design. For example, the revival and transformation of New York City's abandoned railway High Line into a linear park exhibits NBU while revitalizing tourism (Lang & Rothenberg, 2017). Another exemplification can be Seoul's restoration efforts for Cheonggyecheon stream aimed at integrating nature with the urban setting (Lee et al., 2020).

Contrarily, a deeper association between natural capital and the ecological paradigms and procedures with urban ecosystems is envisaged in NDU. Nature is considered to be a primary motivator responsible for acknowledging anthropogenic relationships and urban landscapes (Wu, 2013). By mirroring natural processes as well as encouraging symbiotic interactions between man and nature, NDU aspires to design urban environments that function as ecological systems (Kanaani, 2022). Thus, it lays particular emphasis on developing comprehensive, integrated and interconnected solutions to socio-economic and environmental challenges prevalent in urban settings. Table 2.1 exemplifies the contribution of NDU for the attainment of carbon neutrality.

TABLE 2.1
Examples of nature-driven urbanism

Urban area	NDU interventions
Vauban, Germany	The community emphasizes pedestrian-friendly design, extensive green spaces, and energy-efficient buildings powered by renewables (Omole et al. 2024). Vauban's car-free streets, efficient public transportation and mixed-use development reduce carbon emissions and enhance residents' quality of life (East 2018).
Bilbao Ría 2000, Bilbao, Spain	An urban regeneration project in Spain, revitalized former industrial areas along the Nervión River into vibrant waterfront districts (Alayo et al. 2016). Prioritizing brownfield remediation, ecological restoration, and green infrastructure creation, such as parks and wetlands (Herzog and Rozado 2019), the project enhances biodiversity and contributes to carbon sequestration and climate resilience (Peña et al. 2018). Additionally, it fosters economic revitalization and social cohesion.

(Continued)

TABLE 2.1 (CONTINUED)

Urban area	NDU interventions
Masdar City in Abu Dhabi, United Arab Emirates	It implements sustainable technologies like solar power, wind turbines, and passive cooling systems to cut energy use and greenhouse gas emissions (Omar 2018). Prioritizing walkability, green spaces and water-efficient landscaping, it boosts urban biodiversity and reduces environmental impacts (Juaidi et al. 2019). Masdar City sets a global benchmark for environmentally conscious urban planning and design.
Curitiba, Brazil	The Bus Rapid Transit system, along with pedestrian-friendly design and abundant green spaces, has significantly reduced reliance on private vehicles and resulting carbon emissions (Cervero and Golub 2007). The city's commitment to urban greenery through parks, botanical gardens, and tree-lined streets enhances carbon sequestration, urban biodiversity, and air quality while improving urban microclimates (Cabanek et al. 2020).

Consequently, on assessment NDU is considered to be a more inclusive and complete approach as compared to NBU. NDU takes into account a wider-ranging ecological perspective for creating, designing and planning innately resilient and sustainable urban environments. NDU envisages urban areas that reduce their adverse environmental effects and co-exist harmoniously with nature. To conclude, NDU is a revolutionary approach aimed at revitalizing, reinvigorating and rethinking cities as vibrant, thriving ecosystems which are in a symbiotic relationship with their surrounding natural ecosystems.

REFERENCES

Ackerman, K., Conard, M., Culligan, P., Plunz, R., Sutto, M. P., & Whittinghill, L. (2014). Sustainable food systems for future cities: The potential of urban agriculture. *The Economic and Social Review*, *45*(2, Summer), 189–206.

Adam, P. (2019). Salt marsh restoration. In G. Perillo, E. Wolanski, D. R. Cahoon, & C. S. Hopkinson (Eds.), *Coastal wetlands: An integrated ecosystem approach* (2nd ed., pp. 817–861). Elsevier.

Aerts, R., Honnay, O., & Van Nieuwenhuyse, A. (2018). Biodiversity and human health: Mechanisms and evidence of the positive health effects of diversity in nature and green spaces. *British Medical Bulletin*, 127(1), 5–22.

Akaeze, O., & Nandwani, D. (2020). Urban agriculture in Asia to meet the food production challenges of urbanization: A review. *Urban Agriculture & Regional Food Systems*, 5(1), e20002.

Alayo, J., Henry, G., & Plaza, B. (2016). Bilbao case study. In D. K. Carter (Ed.), *Remaking post-industrial cities: Lessons from North America and Europe* (pp. 137–152), 1st Ed. Routledge.

Alikhani, S., Nummi, P., & Ojala, A. (2021). Urban wetlands: A review on ecological and cultural values. *Water*, 13(22), 3301.

Allen, E., & Clawson, L. (2018). Growing food and farming literacy. *Youth in Food*, *35*, 22–23. https://ruaf.org/assets/2019/11/UA-Magazine-no.-35-Youth-in-Food-Opportunities-for-education-and-employment.pdf#page=21
Amos, C. C., Rahman, A., Karim, F., & Gathenya, J. M. (2018). A scoping review of roof harvested rainwater usage in urban agriculture: Australia and Kenya in focus. *Journal of Cleaner Production*, 202, 174–190.
Angel, S., Parent, J., Civco, D. L., Blei, A., & Potere, D. (2011). The dimensions of global urban expansion: Estimates and projections for all countries, 2000–2050. *Progress in Planning*, 75(2), 53–107. https://doi.org/10.1016/j.progress.2011.04.001
Angelopulo, G. (2021). A comparative measure of inclusive urbanisation in the cities of Africa. *World Development Perspectives*, 22, 100313.
Atkinson, R., & Bridge, G. (2005). *Gentrification in a global context: The new urban colonialism*. Routledge.
Bayulken, B., Huisingh, D., & Fisher, P. M. (2021). How are nature based solutions helping in the greening of cities in the context of crises such as climate change and pandemics? A comprehensive review. *Journal of Cleaner Production*, 288, 125569.
BioRegional (2020). BedZED: The UK's first large-scale eco-village. *BioRegional*. https://www.bioregional.com/projects-and-services/case-studies/bedzed-the-uks-first-large-scale-eco-village
Bratman, G. N., Anderson, C. B., Berman, M. G., Cochran, B., de Vries, S., Flanders, J., Folke, C., Frumkin, H., Gross, J. J., Hartig, T., Kahn, P. H., Kuo, M., Lawler, J. J., Levin, P. S., Lindahl, T., Meyer-Lindenberg, A., Mitchell, R., Ouyang, Z., Roe, J., Scarlett, L., Smith, J. R., van den Bosch, M., Wheeler, B. W., White, M. P., Zheng, H., & Daily, G. C. (2019). Nature and mental health: An ecosystem service perspective. *Science Advances*, 5(7), eaax0903. https://doi.org/10.1126/sciadv.aax0903
Brown, S., Kruger, C., & Subler, S. (2008). Greenhouse gas balance for composting operations. *Journal of Environmental Quality*, 37(4), 1396–1410. https://doi.org/10.2134/jeq2007.0453
Cabanek, A., Zingoni de Baro, M. E., & Newman, P. (2020). Biophilic streets: A design framework for creating multiple urban benefits. *Sustainable Earth*, 3(1), 1–17.
Calthorpe, P., & Fulton, W. (2001). *The regional city: Planning for the end of sprawl*. Island Press.
Caragliu, A., Del Bo, C., & Nijkamp, P. (2011). Smart cities in Europe. *Journal of Urban Technology*, 18(2), 65–82. https://doi.org/10.1080/10630732.2011.601117
Cervero, R., & Golub, A. (2007). Informal transport: A global perspective. *Transport Policy*, 14(6), 445–457. https://doi.org/10.1016/j.tranpol.2007.04.011
Chastas, P., Theodosiou, T., & Bikas, D. (2016). Embodied energy in residential buildings-towards the nearly zero energy building: A literature review. *Building and Environment*, 105, 267–282. https://doi.org/10.1016/j.buildenv.2016.05.040
Chatterjee, A., Chattopadhyay, R. N., Chatterjee, A., & Chattopadhyay, R. N. (2020). Growth of metropolises and megacities with focus on global south. In *Satellite towns in neo-metropolitan development in India* (pp. 1–28). Advances in 21st Century Human Settlements. Springer. https://doi.org/10.1007/978-981-15-1502-6_1
Cheng, Z., Pang, M. S., & Pavlou, P. A. (2020). Mitigating traffic congestion: The role of intelligent transportation systems. *Information Systems Research, 31(3)*, 653–674.
Chetty, R., Hendren, N., Kline, P., & Saez, E. (2014). Where is the land of opportunity? The geography of intergenerational mobility in the United States. *Quarterly Journal of Economics*, 129(4), 1553–1623. https://doi.org/10.1093/qje/qju022
City of Los Angeles Departments of Transportation and City Planning, T. D. M. (2021). Transportation demand management in land use development program. *Transportation,* 1(1), 1–42.
Clarivate (2022). *Web of science: Steps to analyze results*. Clarivate Scientific and Academic Research.

CoDyre, M., Fraser, E. D. G., & Landman, K. (2015). How does your garden grow? An empirical evaluation of the costs and potential of urban gardening. *Urban Forestry and Urban Greening*, 14(1), 72–79. https://doi.org/10.1016/j.ufug.2014.11.001

Crabill, D. (2009). Project for public spaces: Could a change in message facilitate better public space planning? *Agora Journal of Urban Planning and Design*, *3*, 43–47. https://deepblue.lib.umich.edu/handle/2027.42/120343

de Sousa Santos, B. (1998). Participatory budgeting in Porto Alegre: Toward a redistributive democracy. *Politics and Society*, 26(4), 461–510. https://doi.org/10.1177/0032329298026004003

Donthu, N., Kumar, S., Mukherjee, D., Pandey, N., & Lim, W. M. (2021). How to conduct a bibliometric analysis: An overview and guidelines. *Journal of Business Research*, 133, 285–296. https://doi.org/10.1016/j.jbusres.2021.04.070

Duany, A., Plater-Zyberk, E., & Speck, J. (2000). *Suburban nation: The rise of sprawl and the decline of the American dream*. North Point Press.

East, M. (2018). Rising cities: Continuity, innovation and deliberation of Vauban district. *Ecocycles*, 4(2), 58–71.

Ellen MacArthur Foundation (2020). *The circular economy: A transformative Covid-19 recovery strategy - how policymakers can pave the way to a low carbon, prosperous future*. Ellen MacArthur Foundation.

EPA (2020). *Advancing sustainable materials management: 2018 fact sheet*. United States Environmental Protection Agency. Retrieved from https://www.epa.gov/sites/default/files/2021-01/documents/2018_ff_fact_sheet_dec_2020_fnl_508.pdf

Ferrini, F., Fini, A., Mori, J., & Gori, A. (2020). Role of vegetation as a mitigating factor in the urban context. *Sustainability*, 12(10), 4247.

Fink, J. H. (2019). Contrasting governance learning processes of climate-leading and-lagging cities: Portland, Oregon, and Phoenix, Arizona, USA. *Journal of Environmental Policy & Planning*, 21(1), 16–29.

Florida, R. (2002). *The rise of the creative class: And how It's transforming work, Leisure,* Community and Everyday Life. Basic Books.

Foley, J. A., DeFries, R., Asner, G. P., Barford, C., Bonan, G., Carpenter, S. R., Chapin, F. S., Coe, M. T., Daily, G. C., Gibbs, H. K., Helkowski, J. H., Holloway, T., Howard, E. A., Kucharik, C. J., Monfreda, C., Patz, J. A., Prentice, I. C., Ramankutty, N., & Snyder, P. K. (2005). Global consequences of land use. *Science*, 309(5734), 570–574. https://doi.org/10.1126/science.1111772

Forster, P. M., Smith, C., Walsh, T., Lamb, W. F., Lamboll, R., Hall, B., … & Zhai, P. (2024). Indicators of global climate change 2023: Annual update of key indicators of the state of the climate system and human influence. *Earth System Science Data*, 16(6), 2625–2658.

Friendly, A. (2013). The right to the city: Theory and practice in Brazil. *Planning Theory and Practice*, 14(2), 158–179. https://doi.org/10.1080/14649357.2013.783098

Gao, T., Liu, F., Wang, Y., Mu, S., & Qiu, L. (2020). Reduction of atmospheric suspended particulate matter concentration and influencing factors of green space in urban forest park. *Forests*, 11(9), 950.

Glaeser, E. L. (2011). *Triumph of the city: How our greatest invention makes us richer, smarter, greener, healthier, and happier*. Penguin Books.

Gössling, S., Choi, A., Dekker, K., & Metzler, D. (2019). The social cost of automobility, cycling and walking in the European Union. *Ecological Economics*, 158, 65–74. https://doi.org/10.1016/j.ecolecon.2018.12.016

Grimm, N. B., Faeth, S. H., Golubiewski, N. E., Redman, C. L., Wu, J., Bai, X., & Briggs, J. M. (2008). Global change and the ecology of cities. *Science*, 319(5864), 756–760. https://doi.org/10.1126/science.1150195

Hall, P. (2014). *Cities of tomorrow: An intellectual history of urban planning and design since 1880*, 4th Ed. Wiley Blackwell. 640p.

Herzog, C., & Rozado, C. A. (2019). The EU–Brazil sector dialogue on nature-based solutions. *Contribution to A Brazilian roadmap on nature-based solutions for resilient cities*. European Union.

Hobsbawm, E. J. (1999). *Industry and empire: From 1750 to the present day*. New Press.

Hollands, R. G. (2008). Will the real smart city please stand up? Intelligent, progressive or entrepreneurial? *City*, 12(3), 303–320. https://doi.org/10.1080/13604810802479126

Huang, K., Li, X., Liu, X., & Seto, K. C. (2019). Projecting global urban land expansion and heat island intensification through 2050. *Environmental Research Letters*, 14(11), 114037.

IEA (2022), World Energy Outlook 2022. *International Energy Agency (IEA), Paris* https://www.iea.org/reports/world-energy-outlook-2022

IPCC (2018). Global warming of 1.5°C. *Intergovernmental Panel on Climate Change (IPCC)* Retrieved from https://www.ipcc.ch/sr15/

International Renewable Energy Agency (IRENA) (2021). *Renewable Energy and Jobs* [Annual review], Vol. 2021. Retrieved from https://www.irena.org/Publications/2021/Oct/Renewable-Energy-and-Jobs-Annual-Review-2021

Jacobs, J. (1961). *The death and life of great American cities*. Random House LLC.

Japan Ministry of the Environment (2020). Annual Report on the Environment, the Sound Material-Cycle Society and Biodiversity in Japan 2020. *Ministry of the Environment, Environmental Strategy Division, Minister's Secretariat.* https://dl.ndl.go.jp/pid/12934419

Jennings, V., & Bamkole, O. (2019). The relationship between social cohesion and urban green space: An avenue for health promotion. *International Journal of Environmental Research and Public Health*, 16(3), 452.

Jones, P. (2023). *Human rights and development*. Taylor & Francis.

Jones, L., Anderson, S., Læssøe, J., Banzhaf, E., Jensen, A., Bird, D. N., … & Zandersen, M. (2022). A typology for urban green infrastructure to guide multifunctional planning of nature-based solutions. *Nature-Based Solutions*, 2, 100041.

Juaidi, A., AlFaris, F., Saeed, F., Salmeron-Manzano, E., & Manzano-Agugliaro, F. (2019). Urban design to achieving the sustainable energy of residential neighbourhoods in arid climate. *Journal of Cleaner Production*, 228, 135–152.

Kabisch, N., Frantzeskaki, N., & Hansen, R. (2022). Principles for urban nature-based solutions. *Ambio*, 51(6), 1388–1401.

Kanaani, M. (Ed.). (2022). *The Routledge companion to ecological design thinking: Healthful ecotopian visions for architecture and urbanism*. Taylor & Francis.

Kennedy, C., Cuddihy, J., & Engel-Yan, J. (2015). The changing metabolism of cities. *Journal of Industrial Ecology*, 11(2), 43–59. https://doi.org/10.1162/jie.2007.1107

King, A. (2015). *Global cities*. Routledge.

King, R., Orloff, M., Virsilas, T., & Pande, T. (2017). *Confronting the urban housing crisis in the global south: Adequate, secure, and affordable housing*. World Resources Institute.

Klinenberg, E. (2018). *Palaces for the people: How to build a more equal and united society*. Random House.

Kulak, M., Graves, A., & Chatterton, J. (2013). Reducing greenhouse gas emissions with urban agriculture: A life cycle assessment perspective. *Landscape and Urban Planning*, 111, 68–78. https://doi.org/10.1016/j.landurbplan.2012.11.007

Lang, S., & Rothenberg, J. (2017). Neoliberal urbanism, public space, and the greening of the growth machine: New York City's High Line Park. *Environment and Planning A: Economy and Space*, 49(8), 1743–1761.

Langergraber, G., Pucher, B., Simperler, L., Kisser, J., Katsou, E., Buehler, D., … & Atanasova, N. (2020). Implementing nature-based solutions for creating a resourceful circular city. *Blue-Green Systems*, 2(1), 173–185.

Le Corbusier. (1967). *The radiant city: Elements of a doctrine of urbanism to be used as the basis of our machine-age civilization*. Orion Book Company. 348 pp.

Lee, C. S., Lee, H., Kim, A. R., Pi, J. H., Bae, Y. J., Choi, J. K., … & Moon, J. S. (2020). Ecological effects of daylighting and plant reintroduction to the Cheonggye Stream in Seoul, Korea. *Ecological Engineering*, 152, 105879.

Lees, L. (2008). Gentrification and social mixing: Towards an inclusive urban renaissance? *Urban Studies*, 45(12), 2449–2470. https://doi.org/10.1177/0042098008097099

Mahmoud, I. H., Morello, E., Rizzi, D., & Wilk, B. (2023). Localizing Sustainable Development Goals(SDGs)Throughco-creationofNature-BasedSolutions(NBS)towardsanassessment framework for local governments. In R. Brears (Ed.), *The Palgrave encyclopedia of urban and regional futures* (pp. 980–996). Palgrave Macmillan. https://doi.org/10.1007/978-3-030-51812-7_354-1

Mang, N. S. (2019). Toward a regenerative psychology of place. In K. S. Sagendorf & E. A. Wilkerson (Eds.), *Regenerative urban development, climate change and the common good* (pp. 49–76). Routledge.

Manisalidis, I., Stavropoulou, E., Stavropoulos, A., & Bezirtzoglou, E. (2020). Environmental and health impacts of air pollution: A review. *Frontiers in Public Health*, 8, 14.

McDonald, R. I., Kareiva, P., & Forman, R. T. T. (2008). The implications of current and future urbanization for global protected areas and biodiversity conservation. *Biological Conservation*, 141(6), 1695–1703. https://doi.org/10.1016/j.biocon.2008.04.025

McPhearson, T., Pickett, S. T. A., Grimm, N. B., Niemelä, J., Alberti, M., Elmqvist, T., Weber, C., Haase, D., Breuste, J., & Qureshi, S. (2016). Advancing urban ecology toward a science of cities. *BioScience*, 66(3), 198–212. https://doi.org/10.1093/biosci/biw002

Mejia-Arbelaez, C., Sarmiento, O. L., Mora Vega, R., Flores Castillo, M., Truffello, R., Martínez, L., … & Delclòs-Alió, X. (2021). Social inclusion and physical activity in Ciclovía Recreativa programs in Latin America. *International Journal of Environmental Research and Public Health,* 18(2), 655.

Miles, R., & Song, Y. (2009). "Good" neighborhoods in Portland, Oregon: Focus on both social and physical environments. *Journal of Urban Affairs*, 31(4), 491–509. https://doi.org/10.1111/j.1467-9906.2009.00457.x

Mitsch, W. J., & Gosselink, J. G. (2015). *Wetlands*. 5th Ed. John Wiley & Sons.

Nižetić, S., Djilali, N., Papadopoulos, A., & Rodrigues, J. J. (2019). Smart technologies for promotion of energy efficiency, utilization of sustainable resources and waste management. *Journal of Cleaner Production*, 231, 565–591.

Nowak, D. J., Hirabayashi, S., Bodine, A., & Greenfield, E. (2014). Tree and forest effects on air quality and human health in the United States. *Environmental Pollution*, 193, 119–129. https://doi.org/10.1016/j.envpol.2014.05.028

Oke, T. R. (1982). The energetic basis of the urban heat island. *Quarterly Journal of the Royal Meteorological Society*, 108(455), 1–24. https://doi.org/10.1002/qj.49710845502

Okvat, H. A., & Zautra, A. J. (2011). Community gardening: A parsimonious path to individual, community, and environmental resilience. *American Journal of Community Psychology*, 47(3–4), 374–387. https://doi.org/10.1007/s10464-010-9404-z

Omar, W. F. (2018). Zero carbon city-Masdar City critical analysis. *BAU Journal-Health and Wellbeing*, 1(3), 1.

Omole, F. O., Olajiga, O. K., & Olatunde, T. M. (2024). Sustainable urban design: A review of eco-friendly building practices and community impact. *Engineering Science & Technology Journal*, 5(3), 1020–1030.

Pandis Iveroth, S., Vernay, A. L., Mulder, K. F., & Brandt, N. (2013). Implications of systems integration at the urban level: The case of Hammarby Sjöstad, Stockholm. *Journal of Cleaner Production*, 48, 220–231. https://doi.org/10.1016/j.jclepro.2012.09.012

Peña, L., Onaindia, M., Fernández de Manuel, B., Ametzaga-Arregi, I., & Casado-Arzuaga, I. (2018). Analysing the synergies and trade-offs between ecosystem services to reorient land use planning in Metropolitan Bilbao (northern Spain). *Sustainability*, 10(12), 4376.

Pode, R. (2020). Organic light emitting diode devices: An energy efficient solid state lighting for applications. *Renewable and Sustainable Energy Reviews*, 133, 110043.

Roggema, R. (2020). *Nature-driven urbanism* (pp. 1–8). Springer International Publishing.

Roggema, R. (2021). From nature-based to nature-driven: Landscape first for the design of Moeder Zernike in Groningen. *Sustainability*, 13(4), 2368.

Roggema, R., Tillie, N., & Keeffe, G. (2023). The ReciproCity: Nature-driven urbanism in Western Sydney. In P. Droege (Ed.), *Urban and regional agriculture* (pp. 457–500). Academic Press.

Romanovych, A. (2022). *Environmental recovery in Ukraine: Assessing damages and forming solutions* (Master's thesis, Padua, IT: Università Degli Studi Di Padova. Dipartimento di Ingegneria Civile, Edile e Ambientale).

Sanga, N. (2024). Low-income housing development in India: Strategies for income mixing and inclusive urban planning. In A. M. Santiago, I. Levin, & K. Arthurson (Eds.), *Creating mixed communities through housing policies* (pp. 35–52). Routledge.

Saunders, L. E., Green, J. M., Petticrew, M. P., Steinbach, R., & Roberts, H. (2013). What are the health benefits of active travel? A systematic review of trials and cohort studies. *PLOS ONE*, 8(8), e69912. https://doi.org/10.1371/journal.pone.0069912

Seto, K. C., Güneralp, B., & Hutyra, L. R. (2012). Global forecasts of urban expansion to 2030 and direct impacts on biodiversity and carbon pools. *Proceedings of the National Academy of Sciences*, 109(40), 16083–16088. https://doi.org/10.1073/pnas.1211658109

Shenzhen Statistics Bureau (2021). *Shenzhen statistical yearbook 2021*. Retrieved from Shenzhen Statistics. needs link https://www.stats.gov.cn/sj/ndsj/2021/indexeh.htm

Singh, N., Singh, S., & Mall, R. K. (2020). Urban ecology and human health: Implications of urban heat island, air pollution and climate change nexus. In P. Verma, P. Singh, R. Singh, & A. S. Raghubanshi (Eds.), *Urban ecology: Emerging patterns and social-ecological systems* (pp. 317–334). Elsevier.

Singh, P., Singh, V., & Singh, R. K. (2023). Emerging trends in green building construction in India: An eco-friendly technology. *Journal of Global Communication, 16*(1), 5–10.

Snep, R. P., Klostermann, J., Lehner, M., & Weppelman, I. (2023). Social housing as focus area for nature-based solutions to strengthen urban resilience and justice: Lessons from practice in the Netherlands. *Environmental Science & Policy,* 145, 164–174.

Song, C. (2022). Application of nature-based measures in China's sponge city initiative: Current trends and perspectives. *Nature-Based Solutions,* 2, 100010.

Sovacool, B. K., Hook, A., Martiskainen, M., Brock, A., & Turnheim, B. (2020). The decarbonisation divide: Contextualizing landscapes of low-carbon exploitation and toxicity in Africa. *Global Environmental Change,* 60, 102028. https://doi.org/10.1016/j.gloenvcha.2019.102028

Spreiregen, P. D. (1965). *Urban design: The architecture of towns and cities*. McGraw-Hill.

Stanley, J., Stanley, J., & Hansen, R. (2023). Economic influences on strategic land use transport policy and planning. In Stanley, J., Stanley, J., & Hansen, R. (Eds.), *How great cities happen : Integrating people, land use and transport*(pp. 52–81). Edward Elgar Publishing.

Sverige, A. (2021). *Swedish waste management 2021*. Swedish Waste Management Association. https://www.avfallsverige.se/

Taylor, D. E., & Ard, K. J. (2015). Food availability and the food desert frame in Detroit: An overview of the city's food system. *Environmental Practice*, 17(2), 102–133.

UN (2016). *World Cities Report: Urbanization and development – emerging futures. UN-Habitat.* United Nations Human Settlements Programme. Retrieved from https://unhabitat.org/sites/default/files/download-manager-files/WCR-2016-WEB.pdf

UN (2018). *68% of the world population projected to live in urban areas by 2050, says UN.* United Nations. Retrieved from https://www.un.org/development/desa/en/news/population/2018-revision-of-world-urbanization-prospects.html

United Nations Environment Programme. (2023). "*Emissions Gap Report 2023: The Closing Window – Climate crisis calls for rapid transformation of societies.*" *UNEP*. Retrieved from https://www.unep.org/emissions-gap-report-2023

Van Eck, N. J., & Waltman, L. (2010). Software survey: VOSviewer, a computer program for bibliometric mapping. *Scientometrics*, 84(2), 523–538. https://doi.org/10.1007/s11192-009-0146-3

Velenturf, A. P., & Purnell, P. (2021). Principles for a sustainable circular economy. *Sustainable Production and Consumption*, 27, 1437–1457.

Verma, P., Kumari, T., & Raghubanshi, A. S. (2021). Energy emissions, consumption and impact of urban households: A review. *Renewable and Sustainable Energy Reviews,* 147, 111210.

Wong, N. H., Tan, C. L., Kolokotsa, D. D., & Takebayashi, H. (2021). Greenery as a mitigation and adaptation strategy to urban heat. *Nature Reviews Earth & Environment,* 2(3), 166–181.

WGBC (2018). *The business case for green building: A review of the costs and benefits for developers, investors and occupants.* World Green Building Council. Retrieved from https://www.worldgbc.org/business-case

WHO (2019). *Air pollution.* World Health Organization. Retrieved from https://www.who.int/health-topics/air-pollution

Wong, N. H., Cheong, D. W., Yan, H., Soh, J., Ong, C. L., & Sia, A. (2003). The effects of rooftop garden on energy consumption of a commercial building in Singapore. *Energy and Buildings*, 35(4), 353–364.

Wu, J. (2013). Landscape sustainability science: Ecosystem services and human well-being in changing landscapes. *Landscape Ecology*, 28, 999–1023.

3 Sustainable Cities

Integrating Nature-Based Principles with Smart Technologies

Agbugui Marian Onwude, Abhulimen Fran Ehizokhale, Inobeme Abel, Okhamafe Elizabeth Princess, Adeniyi Osaimianmionmwan Arasomwan, and Odozie John Chiwuzulum

3.1 INTRODUCTION

3.1.1 Overview of Climate Change Adaptation

Climate change adaptation refers to the process of adjusting natural or human systems in response to actual or expected climatic stimuli or their effects, which moderates harm or exploits beneficial opportunities (Pisor et al., 2023). As climate change accelerates, cities—home to more than half of the global population face escalating risks, including extreme weather events, rising sea levels, and increased temperatures. Urban areas are particularly vulnerable due to their dense populations, critical infrastructure, and economic activities that are sensitive to climate impacts. Adaptation strategies in these environments are essential for safeguarding public health, economic stability, and the built environment.

The urgency of climate change adaptation in urban contexts cannot be overstated. Cities are already experiencing the adverse effects of climate change, with significant implications for their sustainability and resilience (Jabareen, 2013). For instance, coastal cities are threatened by sea-level rise, while inland cities face challenges related to heat waves and water scarcity. Without effective adaptation measures, these urban centres risk severe disruptions. Consequently, integrating adaptation strategies into urban planning has become a priority for ensuring the long-term viability of cities (Aboagye and Sharifi, 2024). These strategies must be comprehensive, encompassing not only physical infrastructure but also social and institutional frameworks that enhance the capacity of urban communities to withstand and recover from climate-related shocks.

DOI: 10.1201/9781003518037-4

3.1.2 Defining Nature-based Solutions

Nature-based solutions (NbS) are defined as actions that work with and enhance nature to help address societal challenges, including climate change, biodiversity loss, and water security (Dunlop et al., 2024). These solutions leverage natural processes and ecosystems to deliver environmental, social, and economic benefits. NbS can take various forms in urban settings, including the creation of green roofs, the restoration of wetlands, and the development of urban forests. By harnessing the power of nature, NbS contribute to climate change adaptation while simultaneously promoting biodiversity and improving the quality of life for urban residents.

In the context of urban sustainability and climate change adaptation, NbS are particularly relevant due to their multifunctional benefits. They provide a cost-effective and flexible approach to enhancing urban resilience against climate impacts (Ferrario et al., 2024). For example, urban green spaces can mitigate the urban heat island effect, reduce flood risks through natural water retention, and enhance air quality. Moreover, NbS offer an inclusive approach to adaptation by engaging communities in the co-creation and stewardship of natural spaces, thereby fostering social cohesion and equity. As cities strive to achieve carbon neutrality and sustainability, integrating NbS into planning and policy frameworks represents a critical pathway for building adaptive capacity and resilience to climate change.

3.1.3 Understanding the Smart City Paradigm

A "smart city" is generally defined as an urban area that leverages digital technology, data analytics, and innovative practices to enhance the quality of life for its inhabitants, improve efficiency in the management of urban services, and promote sustainable development (Mohanty and Kumar, 2021). The concept of a smart city is inherently multidimensional, encompassing various domains such as transportation, energy, healthcare, public safety, and environmental management. The primary objective of a smart city is to create a more efficient, liveable, and resilient urban environment by integrating advanced technologies with infrastructure, governance, and citizen engagement.

Technology, data, and innovation are the core pillars of smart city development. Advanced technologies, such as the Internet of Things (IoT), artificial intelligence (AI), and big data analytics, play a crucial role in collecting and processing vast amounts of data generated within urban environments (Herath and Mittal, 2022). This data-driven approach enables city planners and managers to make informed decisions, optimise resource use, and anticipate challenges before they escalate. For instance, smart grids optimise energy consumption and distribution, reducing waste and enhancing the reliability of energy supply. Similarly, smart transportation systems utilise real-time data to manage traffic flows, reduce congestion, and lower carbon emissions.

Innovation in governance and public participation is also vital in the smart city paradigm. Smart cities are not only about deploying the latest technologies but also about fostering a culture of continuous improvement and inclusivity. This involves engaging citizens in the decision-making process, utilising technology to facilitate communication and collaboration, and ensuring that the benefits of smart city initiatives are equitably distributed across all segments of the population (Viale Pereira et al., 2017). Through these mechanisms, smart cities aim to create environments that are not only technologically advanced but also socially and environmentally sustainable.

3.1.4 Intersection of Smart Cities and Climate Adaptation

Smart cities are uniquely positioned to address the multifaceted challenges posed by climate change. The integration of digital technology and data analytics allows for more precise monitoring of environmental conditions, better prediction of climate-related events, and more effective management of urban systems under stress (Bibri, 2023). For example, smart water management systems can detect leaks in real-time, preventing water loss and ensuring a sustainable supply even in times of drought. Similarly, AI-driven models can predict the impact of extreme weather events, enabling cities to prepare and respond more effectively.

One of the key areas where smart cities intersect with climate adaptation is in the deployment of NbS. The integration of NbS within the smart city framework enhances urban resilience by combining the strengths of natural ecosystems with technological innovation. For instance, smart green infrastructure—such as IoT-enabled green roofs and vertical gardens—can be monitored and managed to optimise their benefits, such as reducing urban heat island effects, enhancing biodiversity, and improving air quality (Lin et al., 2021). By using sensors and data analytics, cities can track the performance of these NbS in real-time, making adjustments as needed to ensure they are functioning optimally under changing climatic conditions.

Furthermore, smart cities can leverage data to prioritise and scale up NbS. For example, data from satellite imagery, drones, and remote sensors can be used to identify areas within a city that are most vulnerable to climate impacts, such as flooding or heatwaves. This information can guide the strategic placement of NbS, ensuring that they deliver maximum benefits where they are needed most (Zhao et al., 2023). In addition, smart city platforms can facilitate the integration of NbS into existing urban infrastructure, such as incorporating permeable pavements into road networks to manage stormwater or retrofitting buildings with green facades to enhance energy efficiency.

The role of innovation in governance is also crucial in the smart city paradigm, particularly in relation to climate adaptation. By fostering collaboration between different sectors—such as urban planning, environmental management, and technology development—smart cities can create more cohesive and comprehensive adaptation strategies. For example, participatory platforms can be used to involve citizens in the design and implementation of NbS, ensuring that these solutions meet the needs of local communities and are supported by the public (Mora et al., 2023). This inclusive approach not only enhances the effectiveness of climate adaptation efforts but also strengthens the social fabric of the city, making it more resilient to future challenges.

3.2 KEY PRINCIPLES OF NbS (CONCEPTS AND APPLICATIONS)

NbS are guided by several fundamental principles that ensure they contribute effectively to environmental sustainability and urban resilience. The first principle is the enhancement of biodiversity and ecosystems. NbS are designed to work with nature, not against it, by restoring and conserving natural habitats, which in turn support biodiversity and ecological processes (Albert et al., 2021). This principle aligns closely with urban sustainability goals by ensuring that city development does not come at the expense of the natural environment.

The second principle is multifunctionality, which emphasises that NbS should deliver multiple benefits simultaneously. For example, a single NbS, such as an urban wetland, can provide flood protection, improve water quality, offer recreational spaces, and enhance local biodiversity (Dorst et al., 2019). This multifunctionality is crucial in urban planning, where land use is often constrained, and solutions need to be efficient and cost-effective.

The third principle is inclusive governance and community involvement. NbS should be designed and implemented in collaboration with local communities, ensuring that the solutions are socially equitable and culturally appropriate (Kabisch et al., 2022). This principle is particularly important in urban planning, where diverse stakeholder interests must be balanced, and where community buy-in is essential for the long-term success of sustainability initiatives.

These principles align with broader sustainability goals in urban planning by promoting resilience, social equity, and environmental stewardship. By adhering to these principles, NbS can help cities adapt to climate change while also enhancing the overall quality of life for urban residents.

3.3 TYPES OF NbS

NbS can be categorised into several types based on their specific functions and applications within urban environments.

1. **Green Infrastructure:** This category includes elements such as green roofs, green walls, and urban green spaces. Green roofs, for example, provide insulation for buildings, reduce urban heat islands, and manage stormwater runoff (Mihalakakou et al., 2023). Urban green spaces, such as parks and community gardens, offer recreational opportunities, improve air quality, and support biodiversity.
2. **Water Management Solutions:** These NbS are designed to manage water resources sustainably and include constructed wetlands, riparian buffers, and permeable pavements. Constructed wetlands, for example, treat wastewater naturally while providing habitat for wildlife (Ferreira et al., 2023; Ganapathi et al., 2024). Permeable pavements allow rainwater to infiltrate the ground, reducing flood risks and recharging aquifers.
3. **Restoration of Natural Ecosystems:** This involves the restoration of degraded ecosystems, such as reforestation, wetland restoration, and the rewilding of urban areas. Restoring wetlands can reduce flood risks,

sequester carbon, and create habitat for species (Cheng and Li, 2024). Reforestation projects in urban areas can improve air quality, sequester carbon, and provide shade.

4. **Coastal Resilience Solutions:** These NbS include the restoration of mangroves, coral reefs, and dunes to protect coastal cities from storm surges and sea-level rise. Mangrove restoration, for instance, not only protects against erosion but also supports fisheries and sequesters carbon (Châles et al., 2023).

By incorporating these diverse types of NbS into urban planning, cities can address a range of environmental challenges, from climate adaptation to biodiversity conservation.

3.4 CASE STUDIES: SUCCESSFUL NBS IMPLEMENTATIONS IN CITIES

1. **Copenhagen, Denmark:** Copenhagen has successfully implemented NbS through its Climate Resilient Neighbourhoods initiative, which focuses on using green infrastructure to manage stormwater and reduce flooding (Brears, 2023). The city installed green roofs, permeable pavements, and green streets, which not only manage water effectively but also enhance the urban environment. The project has significantly reduced the risk of flooding and created more green spaces for residents. A key challenge was coordinating between different municipal departments, but the project highlighted the importance of integrated planning.
2. **Singapore:** Singapore's Bishan-Ang Mo Kio Park is a flagship example of NbS in urban water management. The city transformed a concrete canal into a naturalised river with surrounding parkland, which provides flood protection, improves water quality, and offers recreational space for the community (An et al., 2020). The park has become a popular urban oasis, demonstrating the potential of NbS to combine functionality with public amenities. The project faced initial scepticism regarding the removal of the concrete canal, but the success of the transformation has led to wider acceptance of NbS in urban planning.
3. **New York City, USA:** New York City's Million Trees NYC initiative aimed to plant 1 million trees across the city to enhance urban resilience (Jones and Goodkind, 2019). The trees help to reduce urban heat islands, improve air quality, and manage stormwater. The project has also fostered community engagement and stewardship, with residents involved in tree planting and maintenance. One challenge was ensuring tree survival in the harsh urban environment, which was addressed through careful species selection and long-term maintenance planning.

These case studies illustrate the diverse applications and benefits of NbS in urban contexts. They also highlight common challenges, such as the need for cross-sectoral collaboration and long-term maintenance, which are crucial for the success of NbS projects. The lessons learned from these cities can inform future NbS initiatives, helping to create more resilient and sustainable urban environments.

3.5 TECHNOLOGICAL INTEGRATION FOR NBS IN SMART CITIES

The integration of advanced technologies such as the IoT, AI, and big data analytics plays a pivotal role in enhancing the effectiveness of NbS within smart cities. These technologies facilitate the monitoring, management, and optimisation of NbS, ensuring that they deliver maximum benefits in terms of climate adaptation, biodiversity enhancement, and urban sustainability.

IoT enables real-time monitoring of environmental conditions and the performance of NbS. For instance, IoT sensors can be deployed in green infrastructure, such as urban forests or green roofs, to monitor factors like soil moisture, temperature, and air quality (Chamara et al., 2022). These sensors provide continuous data that can be analysed to optimise the maintenance and performance of these NbS. For example, if the soil moisture in an urban park falls below a certain threshold, an automated irrigation system can be triggered, ensuring that the vegetation remains healthy without wasting water.

AI further enhances the capabilities of NbS by enabling predictive analytics and decision-making support. AI algorithms can process large datasets collected from IoT sensors, weather stations, and satellite imagery to predict environmental changes and assess the potential impacts on NbS (Jain et al., 2023). For example, AI can predict the likelihood of flooding in a specific urban area based on real-time weather data and historical flood patterns. This information can be used to proactively manage urban wetlands or green spaces to absorb excess water, thereby mitigating flood risks.

In addition to monitoring and predictive capabilities, big data analytics allows for the integration of diverse datasets to understand the broader impacts of NbS on urban environments. By analysing data from various sources, including social media, public health records, and environmental sensors, cities can assess the social, economic, and environmental benefits of NbS (Wang et al., 2024). For example, data analytics can reveal how urban green spaces reduce heat-related illnesses or improve mental well-being, providing evidence to support the expansion of NbS in smart city planning.

Examples of tech-enabled NbS monitoring and management systems include the CityTree, an IoT-enabled air filtration system that uses moss cultures to clean urban air, and Smart Green Roofs, which utilise sensors and AI to optimise water use and plant health on building rooftops (Villaluz, 2017). These systems demonstrate the potential of technology to enhance the performance and sustainability of NbS, making them more effective in addressing urban climate challenges.

3.6 POLICY AND GOVERNANCE FRAMEWORKS

Effective policy and governance frameworks are essential for the successful integration of NbS into smart cities. These frameworks provide the regulatory and institutional support needed to implement, manage, and sustain NbS initiatives, ensuring that they contribute to urban resilience and sustainability over the long term.

One key aspect of policy frameworks is the incorporation of NbS into urban planning regulations and standards. Cities like Glasgow and Genk have developed specific guidelines that mandate the inclusion of green infrastructure in new developments, such as requiring a certain percentage of roof space to be covered by vegetation (Frantzeskaki et al., 2020). These policies ensure that NbS are systematically integrated into urban landscapes, contributing to broader climate adaptation and sustainability goals.

Governance plays a critical role in coordinating the various stakeholders involved in NbS projects, including government agencies, private sector partners, and local communities (Frantzeskaki et al., 2020). Effective governance frameworks establish clear roles and responsibilities, streamline decision-making processes, and facilitate collaboration across different sectors. For instance, in the Netherlands, the national government collaborates with municipalities to implement NbS as part of its Delta Programme, which focuses on flood risk management and climate adaptation (Van Alphen, 2016).

Long-term success of NbS also depends on the availability of financial mechanisms and incentives. Policies that provide funding for NbS projects, such as green bonds or tax incentives for green infrastructure investments, can encourage private sector participation and ensure the financial sustainability of these initiatives (Kabisch et al., 2022). Additionally, governance frameworks that promote public-private partnerships can leverage the expertise and resources of both sectors, leading to more innovative and scalable NbS solutions.

Monitoring and evaluation frameworks are also critical for the governance of NbS. These frameworks establish metrics and indicators to assess the performance and impact of NbS over time, allowing cities to make data-driven adjustments and improvements (Sowińska-Świerkosz and García, 2021). For example, performance indicators might include the reduction in surface temperatures due to urban green spaces or the increase in biodiversity as a result of wetland restoration. By regularly monitoring these indicators, cities can ensure that NbS continue to meet their intended objectives and contribute to urban resilience.

3.7 STAKEHOLDER ENGAGEMENT AND PUBLIC PARTICIPATION

The involvement of local communities and stakeholders is crucial for the successful planning, implementation, and maintenance of NbS in smart cities. Engaging stakeholders ensures that NbS initiatives are socially inclusive, culturally appropriate, and responsive to the needs and priorities of the people who live and work in urban areas.

Stakeholder engagement in NbS planning often involves participatory processes that bring together residents, local businesses, environmental organisations, and government agencies to co-design and co-manage NbS projects (Pagano et al., 2019). This approach not only enhances the relevance and effectiveness of NbS but also fosters a sense of ownership and stewardship among community members. This engagement helped build public support for the project and ensured that the urban forest met the needs of different communities.

Successful stakeholder engagement processes also include ongoing education and communication efforts. By raising awareness about the benefits of NbS and providing opportunities for community involvement, cities can cultivate long-term support for these initiatives (McEvoy et al., 2024).

3.8 BARRIERS TO IMPLEMENTING NBS IN SMART CITIES

While NbS offer significant potential for enhancing urban resilience and sustainability, their implementation in smart cities faces several challenges. One of the primary barriers is funding. NbS often require substantial upfront investment, and the benefits, such as improved air quality or reduced flood risk, may take years to materialise. This temporal mismatch between costs and benefits can deter investment, especially in cities with constrained budgets (Kabisch et al., 2022). Additionally, securing consistent and long-term funding for maintenance and monitoring of NbS is challenging, as these needs are often overlooked during the planning stages.

Another significant challenge is policy gaps. In many cities, existing regulations and planning frameworks are not designed to accommodate or incentivise NbS. Traditional urban planning approaches often prioritise grey infrastructure solutions, such as concrete flood barriers, over green infrastructure like wetlands or urban forests (Mohanty, 2020). This bias can be exacerbated by a lack of clear guidelines or standards for NbS, making it difficult for planners to justify their use or integrate them into existing urban development plans.

Technological constraints also pose a barrier, particularly in the context of integrating NbS with smart city technologies. While IoT and AI can enhance the monitoring and management of NbS, these technologies are still evolving and can be costly to implement and maintain. Moreover, the interoperability of different systems—such as integrating environmental sensors with citywide data platforms—can be complex and require significant technical expertise (Bećirović, 2023). These technological challenges can limit the scalability and effectiveness of NbS, especially in cities with limited access to advanced technology or skilled personnel.

3.9 OPPORTUNITIES FOR SCALING UP NbS

Despite these challenges, there are significant opportunities for scaling up NbS in smart cities, particularly through innovation and collaboration. Advances in technology, such as more affordable and efficient IoT devices and AI-driven analytics, are making it easier to monitor and optimise NbS, thereby enhancing their effectiveness and reducing maintenance costs (Nelson et al., 2020). Innovations in financing, such as green bonds and public-private partnerships, also provide new avenues for funding NbS projects. These financial mechanisms can attract investment by offering returns linked to the environmental benefits generated by NbS, such as carbon sequestration or water purification.

Collaboration between public, private, and civil society sectors is another key opportunity. Public-private partnerships can leverage the strengths of both sectors, combining public oversight with private sector efficiency and innovation. For example, private companies can contribute technological expertise and resources, while public agencies ensure that NbS projects align with broader urban sustainability goals (Frantzeskaki et al., 2020). Additionally, engaging local communities in the design and implementation of NbS can enhance their social acceptability and ensure that the solutions are tailored to local needs and conditions.

International cooperation also presents an opportunity for scaling up NbS. Cities can share knowledge, tools, and best practices through global networks, such as the C40 Cities Climate Leadership Group, which supports urban leaders in implementing NbS and other climate adaptation strategies (Kabisch et al., 2022). Such cooperation can help cities overcome common barriers, such as funding and policy gaps, by providing access to shared resources and expertise.

3.10 KEY PERFORMANCE INDICATORS (KPIS) FOR NbS

Key Performance Indicators (KPIs) are essential for assessing the effectiveness of NbS in smart cities. These indicators provide measurable benchmarks that help to evaluate the success of NbS in achieving its intended environmental, social, and economic goals. Relevant KPIs for NbS include biodiversity enhancement, which can be measured through species richness and abundance in urban green spaces (Mabrouk et al., 2023). Climate regulation is another critical KPI, often assessed by monitoring reductions in urban heat islands and greenhouse gas emissions, such as the amount of carbon sequestered by urban forests (Kabisch et al., 2022). Water management effectiveness can be evaluated by measuring changes in stormwater runoff volume and water quality improvements due to green infrastructure like wetlands and permeable pavements (Wang et al., 2023). Additionally, social well-being KPIs, such as increased public use of green spaces and improved mental health outcomes, can gauge the broader societal impacts of NbS (Frantzeskaki et al., 2020).

3.11 LONG-TERM MONITORING STRATEGIES

Long-term monitoring and data collection are crucial for the sustained success of NbS. Continuous monitoring allows cities to assess the ongoing performance of NbS and make necessary adjustments to optimise their benefits. For example, monitoring soil moisture levels in urban green spaces ensures that vegetation remains healthy and continues to provide ecosystem services, such as cooling and carbon sequestration (Li et al., 2023). Long-term data collection also helps in understanding the evolving impacts of NbS under changing climatic conditions, allowing for adaptive management practices (Kabisch et al., 2022). Moreover, sustained monitoring supports the accumulation of data over time, which can be used to refine predictive models and improve future NbS planning and implementation (Mabrouk et al., 2023). By integrating long-term monitoring strategies into NbS projects, cities can ensure that these solutions continue to deliver resilience and sustainability benefits well into the future.

3.12 FUTURE RESEARCH AND POLICY RECOMMENDATIONS

Future research should focus on developing standardised metrics for evaluating the effectiveness of NbS across different urban contexts. This includes refining KPIs and improving methodologies for long-term monitoring and data collection. Additionally, more research is needed to understand the socio-economic impacts of

NbS, particularly in terms of equity and social inclusion, to ensure that these solutions benefit all urban residents equitably.

For policymakers and urban planners, it is essential to integrate NbS into existing urban planning frameworks and to develop policies that incentivise the use of green infrastructure over traditional grey solutions. This includes revising zoning laws, offering financial incentives for NbS projects, and promoting public-private partnerships that leverage the strengths of both sectors. Furthermore, community engagement should be prioritised in all NbS initiatives to ensure that they are responsive to local needs and conditions, fostering long-term support and sustainability. By addressing these areas, cities can better position themselves to meet the challenges of climate change while enhancing urban liveability and sustainability.

3.13 CONCLUSION

This chapter has explored the critical role of NbS in enhancing urban sustainability and resilience within the framework of smart cities. We began by outlining the fundamental principles that guide NbS, including the enhancement of biodiversity, multifunctionality, and inclusive governance. These principles align closely with the overarching goals of urban sustainability, providing a foundation for integrating NbS into city planning and development. We then categorised various types of NbS, such as green infrastructure, water management solutions, restoration of natural ecosystems, and coastal resilience solutions, highlighting their diverse applications in addressing urban climate challenges.

Case studies from cities like Copenhagen, Singapore, and New York City demonstrated the practical implementation of NbS, showcasing both the successes and challenges encountered. These examples underscore the importance of integrated planning, community engagement, and sustained investment in ensuring the long-term effectiveness of NbS. Moreover, the discussion on integrating NbS into smart cities emphasised the role of technology—particularly IoT, AI, and big data analytics—in monitoring and optimising NbS. These technologies not only enhance the efficiency of NbS but also facilitate real-time adaptation to changing environmental conditions.

This chapter also addressed the challenges associated with implementing NbS, including funding constraints, policy gaps, and technological limitations. However, it also identified significant opportunities for scaling up NbS, particularly through innovation in financing mechanisms and fostering collaboration between public and private sectors. The importance of long-term monitoring and the establishment of robust policy frameworks were highlighted as critical factors for the sustained success of NbS in urban environments.

In summary, the integration of NbS into smart cities represents a powerful approach to addressing the multifaceted challenges of climate change while simultaneously promoting urban sustainability. By leveraging technological innovations, fostering inclusive governance, and ensuring sustained investment, cities can enhance their resilience and improve the quality of life for their residents.

REFERENCES

Aboagye, P. D., & Sharifi, A. (2024). Urban climate adaptation and mitigation action plans: A critical review. *Renewable and Sustainable Energy Reviews*, 189, 113886. https://doi.org/10.1016/j.rser.2023.113886

Albert, C., Brillinger, M., Guerrero, P., Gottwald, S., Henze, J., Schmidt, S., Ott, E., & Schröter, B. (2021). Planning nature-based solutions: Principles, steps, and insights. *Ambio*, 50(8), 1446–1461. https://doi.org/10.1007/s13280-020-01365-1

An, Z., Chen, Q., & Li, J. (2020). *Ecological strategies of urban ecological parks – A case of Bishan Ang Mo Kio Park and Kallang River in Singapore.* E3S Web of Conferences, 194, 05060. https://doi.org/10.1051/e3sconf/202019405060

Bećirović, S. (2023). Challenges and barriers for effective integration of technologies into teaching and learning. In S. Bećirović (Eds.), *Digital pedagogy* (pp. 123–133). Springer Nature Singapore. https://doi.org/10.1007/978-981-99-0444-0_10

Bibri, S. E. (2023). Data-driven smart eco-cities of the future: An empirically informed integrated model for strategic sustainable urban development. *World Futures*, 79(7–8), 703–746. https://doi.org/10.1080/02604027.2021.1969877

Brears, R. C. (2023). Copenhagen becoming a blue-green city through blue-green infrastructure. In R. C. Brears (Ed.), *Blue and green cities* (pp. 115–132). Springer International Publishing. https://doi.org/10.1007/978-3-031-41393-3_6

Châles, F., Bellanger, M., Bailly, D., Dutra, L. X. C., & Pendleton, L. (2023). Using standards for coastal nature-based solutions in climate commitments: Applying the IUCN Global Standard to the case of Pacific Small Island Developing States. *Nature-Based Solutions*, 3, 100034. https://doi.org/10.1016/j.nbsj.2022.100034

Chamara, N., Islam, M. D., Bai, G. (Frank), Shi, Y., & Ge, Y. (2022). Ag-IoT for crop and environment monitoring: Past, present, and future. *Agricultural Systems*, 203, 103497. https://doi.org/10.1016/j.agsy.2022.103497

Cheng, C., & Li, F. (2024). Ecosystem restoration and management based on nature-based solutions in China: Research progress and representative practices. *Nature-Based Solutions*, 100176. https://doi.org/10.1016/j.nbsj.2024.100176

Dorst, H., Van Der Jagt, A., Raven, R., & Runhaar, H. (2019). Urban greening through nature-based solutions – Key characteristics of an emerging concept. *Sustainable Cities and Society*, 49, 101620. https://doi.org/10.1016/j.scs.2019.101620

Dunlop, T., Khojasteh, D., Cohen-Shacham, E., Glamore, W., Haghani, M., Van Den Bosch, M., Rizzi, D., Greve, P., & Felder, S. (2024). The evolution and future of research on nature-based solutions to address societal challenges. *Communications Earth & Environment*, 5(1), 132. https://doi.org/10.1038/s43247-024-01308-8

Ferrario, F., Mourato, J. M., Rodrigues, M. S., & Dias, L. F. (2024). Evaluating nature-based solutions as urban resilience and climate adaptation tools: A meta-analysis of their benefits on heatwaves and floods. *Science of the Total Environment*, 950, 175179. https://doi.org/10.1016/j.scitotenv.2024.175179

Ferreira, C. S. S., Kašanin-Grubin, M., Solomun, M. K., Sushkova, S., Minkina, T., Zhao, W., & Kalantari, Z. (2023). Wetlands as nature-based solutions for water management in different environments. *Current Opinion in Environmental Science & Health*, 33, 100476. https://doi.org/10.1016/j.coesh.2023.100476

Frantzeskaki, N., Vandergert, P., Connop, S., Schipper, K., Zwierzchowska, I., Collier, M., & Lodder, M. (2020). Examining the policy needs for implementing nature-based solutions in cities: Findings from city-wide transdisciplinary experiences in Glasgow (UK), Genk (Belgium) and Poznań (Poland). *Land Use Policy*, *96*, 104688. https://doi.org/10.1016/j.landusepol.2020.104688

Ganapathi, H., Awasthi, S., & Vasudevan, P. (2024). Wetlands as a nature-based solution for urban water management. In A. Stefanakis, H. V. Oral, C. Calheiros, & P. Carvalho (Eds.), *Nature-based solutions for circular management of urban water* (pp. 259–275). Springer International Publishing. https://doi.org/10.1007/978-3-031-50725-0_15

Herath, H. M. K. K. M. B., & Mittal, M. (2022). Adoption of artificial intelligence in smart cities: A comprehensive review. *International Journal of Information Management Data Insights*, 2(1), 100076. https://doi.org/10.1016/j.jjimei.2022.100076

Jabareen, Y. (2013). Planning the resilient city: Concepts and strategies for coping with climate change and environmental risk. *Cities*, 31, 220–229. https://doi.org/10.1016/j.cities.2012.05.004

Jain, H., Dhupper, R., Shrivastava, A., Kumar, D., & Kumari, M. (2023). AI-enabled strategies for climate change adaptation: Protecting communities, infrastructure, and businesses from the impacts of climate change. *Computational Urban Science*, 3(1), 25. https://doi.org/10.1007/s43762-023-00100-2

Jones, B. A., & Goodkind, A. L. (2019). Urban afforestation and infant health: Evidence from MillionTreesNYC. *Journal of Environmental Economics and Management*, 95, 26–44. https://doi.org/10.1016/j.jeem.2019.03.002

Kabisch, N., Frantzeskaki, N., & Hansen, R. (2022). Principles for urban nature-based solutions. *Ambio*, 51(6), 1388–1401. https://doi.org/10.1007/s13280-021-01685-w

Li, Y., Ren, C., Ho, J. Y., & Shi, Y. (2023). Landscape metrics in assessing how the configuration of urban green spaces affects their cooling effect: A systematic review of empirical studies. *Landscape and Urban Planning*, 239, 104842. https://doi.org/10.1016/j.landurbplan.2023.104842

Lin, B. B., Ossola, A., Alberti, M., Andersson, E., Bai, X., Dobbs, C., Elmqvist, T., Evans, K. L., Frantzeskaki, N., Fuller, R. A., Gaston, K. J., Haase, D., Jim, C. Y., Konijnendijk, C., Nagendra, H., Niemelä, J., McPhearson, T., Moomaw, W. R., Parnell, S., … Tan, P. Y. (2021). Integrating solutions to adapt cities for climate change. *The Lancet Planetary Health*, 5(7), e479–e486. https://doi.org/10.1016/S2542-5196(21)00135-2

Mabrouk, M., Han, H., Fan, C., Abdrabo, K. I., Shen, G., Saber, M., Kantoush, S. A., & Sumi, T. (2023). Assessing the effectiveness of nature-based solutions-strengthened urban planning mechanisms in forming flood-resilient cities. *Journal of Environmental Management*, 344, 118260. https://doi.org/10.1016/j.jenvman.2023.118260

McEvoy, D., Tara, A., Vahanvati, M., Ho, S., Gordon, K., Trundle, A., Rachman, C., & Qomariyah, Y. (2024). Localized nature-based solutions for enhanced climate resilience and community wellbeing in urban informal settlements. *Climate and Development*, 16(7), 600–612. https://doi.org/10.1080/17565529.2023.2277248

Mihalakakou, G., Souliotis, M., Papadaki, M., Menounou, P., Dimopoulos, P., Kolokotsa, D., Paravantis, J. A., Tsangrassoulis, A., Panaras, G., Giannakopoulos, E., & Papaefthimiou, S. (2023). Green roofs as a nature-based solution for improving urban sustainability: Progress and perspectives. *Renewable and Sustainable Energy Reviews*, 180, 113306. https://doi.org/10.1016/j.rser.2023.113306

Mohanty, M. (2020). Sustainable urban planning and making sustainable cities. In W. Leal Filho, A. M. Azul, L. Brandli, P. G. Özuyar, & T. Wall (Eds.), *Sustainable cities and communities* (pp. 1–12). Springer International Publishing. https://doi.org/10.1007/978-3-319-71061-7_62-1

Mohanty, R., & Kumar, B. P. (2021). Urbanization and smart cities. In J. R. Vacca (Ed.), *Solving urban infrastructure problems using smart city technologies* (pp. 143–158). Elsevier. https://doi.org/10.1016/B978-0-12-816816-5.00007-3

Mora, L., Gerli, P., Ardito, L., & Messeni Petruzzelli, A. (2023). Smart city governance from an innovation management perspective: Theoretical framing, review of current practices, and future research agenda. *Technovation*, 123, 102717. https://doi.org/10.1016/j.technovation.2023.102717

Nelson, D. R., Bledsoe, B. P., Ferreira, S., & Nibbelink, N. P. (2020). Challenges to realizing the potential of nature-based solutions. *Current Opinion in Environmental Sustainability*, 45, 49–55. https://doi.org/10.1016/j.cosust.2020.09.001

Pagano, A., Pluchinotta, I., Pengal, P., Cokan, B., & Giordano, R. (2019). Engaging stakeholders in the assessment of NBS effectiveness in flood risk reduction: A participatory System Dynamics Model for benefits and co-benefits evaluation. *Science of the Total Environment*, 690, 543–555. https://doi.org/10.1016/j.scitotenv.2019.07.059

Pisor, A. C., Touma, D., Singh, D., & Jones, J. H. (2023). To understand climate change adaptation, we must characterize climate variability: Here's how. *One Earth*, 6(12), 1665–1676. https://doi.org/10.1016/j.oneear.2023.11.005

Sowińska-Świerkosz, B., & García, J. (2021). A new evaluation framework for nature-based solutions (NBS) projects based on the application of performance questions and indicators approach. *Science of the Total Environment*, 787, 147615. https://doi.org/10.1016/j.scitotenv.2021.147615

Van Alphen, J. (2016). The Delta programme and updated flood risk management policies in the Netherlands. *Journal of Flood Risk Management*, 9(4), 310–319. https://doi.org/10.1111/jfr3.12183

Viale Pereira, G., Cunha, M. A., Lampoltshammer, T. J., Parycek, P., & Testa, M. G. (2017). Increasing collaboration and participation in smart city governance: A cross-case analysis of smart city initiatives. *Information Technology for Development*, 23(3), 526–553. https://doi.org/10.1080/02681102.2017.1353946

Villaluz, K. (2017). *This Moss-Covered Air Purifier Can Do the Work of 275 Urban Trees*. Interesting Engineering. Retrieved 1 September 2024, from https://interestingengineering.com/innovation/moss-covered-air-purifier-can-work-of-275-urban-trees

Wang, M., Sun, C., & Zhang, D. (2023). Opportunities and challenges in green stormwater infrastructure (GSI): A comprehensive and bibliometric review of ecosystem services from 2000 to 2021. *Environmental Research*, 236, 116701. https://doi.org/10.1016/j.envres.2023.116701

Wang, N., Chen, B., Wang, L., Ma, Z., & Pan, S. (2024). Big data analytics capability and social innovation: The mediating role of knowledge exploration and exploitation. *Humanities and Social Sciences Communications*, 11(1), 864. https://doi.org/10.1057/s41599-024-03288-8

Zhao, S., Liu, M., Tao, M., Zhou, W., Lu, X., Xiong, Y., Li, F., & Wang, Q. (2023). The role of satellite remote sensing in mitigating and adapting to global climate change. *Science of the Total Environment*, 904, 166820. https://doi.org/10.1016/j.scitotenv.2023.166820

Section II

Green Urbanism

Impact, Adaptation and Mitigation

4 Nature-Based Solutions for Climate Change Adaptation in a Smart City Paradigm in Climate-Related Disasters and Urban Resilience

Chukwuemeka Kingsley John, Fidelis Odedishemi Ajibade, Temitope Fausat Ajibade, Pankaj Kumar, Bashir Adelodun, Adamu Yunusa Ugya, Mathieu Nsenga Kumwimba, Kayode Hassan Lasisi, and James Rotimi Adewumi

4.1 INTRODUCTION

Global urbanization is accelerating, making cities the epicenter of vulnerability and opportunity. More than half of the world's population currently lives in urban regions, which promotes innovation, economic growth, and cross-cultural interaction. But there are also serious drawbacks to this fast urbanization, especially considering climate change. Urban ecosystems face existential risks due to the rising frequency and severity of climate-related calamities such as heat waves, floods, storms, and droughts. In addition to interfering with day-to-day activities, these calamities also put a strain on infrastructure, amplify social inequality, and jeopardize economic stability. In this situation, cities need to come up with creative solutions to deal with these issues and strengthen their defenses against potential threats (Sukanya & Tantia, 2023; Chowdhuri & Gupta, 2023). Nature-based solutions (NbS) can help achieve carbon neutrality by lowering reliance on energy-intensive climate adaptation techniques like water treatment, flood control systems, and mechanical cooling. For example, green roofs and urban forests naturally control temperatures, reducing the need for air conditioning and so lowering energy use and related carbon emissions. By managing rainfall more effectively than conventional drainage infrastructure,

DOI: 10.1201/9781003518037-6

wetlands and permeable surfaces provide improved carbon sequestration while lowering the energy required for pumping and wastewater treatment (Carvalho et al., 2022). The application of NbS has gained traction as a viable strategy for tackling the intricate problems brought about by urban climate change. NbS are tactics that use the power of nature to improve urban resilience, lessen the effects of climate-related calamities, and advance sustainable objectives. Natural background sound provides a comprehensive and sustainable substitute for conventional manufactured solutions by utilizing ecosystems and natural processes. For example, reclaiming wetlands can lower the risk of flooding, reforestation can lessen the impact of heat waves, and green roofs can control stormwater. These actions also provide co-benefits for locals, such as better air quality, biodiversity preservation, and an improved standard of living (Tozer et al., 2022a; Pandey & Ghosh, 2023). The new paradigm of smart cities, which optimize urban systems and raise resident quality of life via the use of digital technologies and data-driven methodologies, is strongly aligned with the idea of NbS. To improve the effectiveness of urban services, lower resource consumption, and advance sustainability, smart cities include cutting-edge technology like the Internet of Things (IoT), artificial intelligence (AI), and big data analytics. The efficiency of climate adaptation measures can be greatly increased when smart city projects are integrated with NbS. AI, for instance, can optimize the location and design of green infrastructure to maximize its benefits, while IoT sensors can monitor environmental conditions in real-time, enabling more precise and adaptable management of NbS (Bibri, 2022; Alahi et al., 2023).

Another important aspect to consider while discussing NbS in smart cities is urban resilience. The ability of a city to endure and rebound from unfavorable occurrences, such as catastrophes linked to climate change, is referred to as resilience. Because they act as organic shock and stress absorbers, NbS are essential for improving urban resilience. Urban parks, for example, can act as cooling areas during heatwaves, and mangroves can shield coastal communities from storm surges. NbS also promotes social resilience by offering areas for leisure and community engagement, which can enhance cohesiveness and aid in times of need. Not only do NbS mitigate current climate threats, but they also advance the long-term objective of urban sustainability, especially when viewed through the prism of carbon neutrality. To achieve carbon neutrality, one must offset or sequester an equal quantity of carbon dioxide emissions to bring net carbon emissions down to zero (Petit, 2022; Cong et al., 2023). By taking up carbon dioxide through activities like photosynthesis and soil carbon storage, NbS serve as carbon sinks and help achieve carbon neutrality. Green infrastructure, such as urban forests and green roofs, not only stores carbon but also lowers energy use by controlling urban temperatures, which lessens the need for heating and cooling buildings. NbS integration is not without difficulties, nevertheless, when it comes to smart city and urban planning frameworks. The local ecological, social, and economic circumstances must be carefully considered when implementing NbS. Moreover, NbS benefits tend to be long-term in nature, making it challenging to time them to coincide with short-term political and economic cycles. Governance issues also arise since NbS frequently calls for cooperation between several sectors and stakeholders, such as local communities, commercial developers, and governmental organizations. Another crucial issue is ensuring that everyone has equal access to the advantages of NbS, especially in urban areas with significant socioeconomic inequality (Pan et al., 2023; He, 2022; Dorst et al., 2022; Farina et al., 2024).

Creating strong governance frameworks and legislative frameworks that facilitate NbS inclusion into smart city programs is crucial to addressing these issues. This entails establishing incentives for NbS adoption, making certain that they are included in laws governing urban planning and development, and encouraging cooperation between stakeholders. In addition, further investigation and creativity are required to enhance the planning, execution, and oversight of NbS, especially given the swiftly advancing landscape of smart city technology. Incorporating NbS into the smart city framework is a viable approach for cities to meet their sustainability objectives, improve urban resilience, and adjust to climate change—especially when it comes to being carbon neutral. Building resilient, sustainable, and habitable urban environments will require a combination of technological and natural solutions as cities continue to grow and face more complicated difficulties. Furthermore, considering how quickly smart city technology is developing, more research and innovation are needed to improve NbS planning, implementation, and supervision. A workable strategy for cities to achieve their sustainability goals, boost urban resilience, and adapt to climate change—especially in terms of becoming carbon neutral—is to integrate NbS within the framework of smart cities. To create resilient, sustainable, and livable urban settings, a blend of technological and natural solutions will be needed as cities continue to grow and encounter increasingly complex problems (Petit, 2022; Castelo et al., 2023).

NbS, which use natural processes to absorb carbon dioxide, improve energy efficiency, and lower emissions, are essential to reaching carbon neutrality. By using photosynthesis to absorb and store atmospheric carbon dioxide (CO), urban woods, for instance, function as efficient carbon sinks (Liu et al., 2023). According to a study by Domke et al. (2022), urban trees in the United States are predicted to absorb over 25.6 million metric tons of carbon yearly, which greatly offsets emissions from urban activities. Urban forests not only sequester carbon but also mitigate the urban heat island (UHI) effect, which reduces the need for air conditioning and, consequently, energy consumption and related greenhouse gas emissions. To further contribute to carbon neutrality, tree canopies also absorb rainwater, lowering stormwater runoff and the energy needed for wastewater treatment (Wang et al., 2021). Another example of NbS that promotes carbon neutrality is green roofs, which lower emissions and increase building energy efficiency. By acting as natural insulation, these vegetative roofing systems assist in controlling indoor temperatures and lessen the need for heating and cooling equipment (Liu et al., 2023; Aljashaami et al., 2024). According to research by Bevilacqua (2021), green roofs can cut summertime air conditioning energy use by up to 75%, which lowers the demand for electricity and lessens the need for fossil fuel-based power generation. Additionally, by enhancing air quality, absorbing particulate matter, and prolonging the life of roofing materials, green roofs help to indirectly reduce carbon emissions from their manufacture and replacement. NbS can greatly contribute to the shift to carbon neutrality while boosting urban resilience by including urban trees and green roofs in city development (Tan et al., 2023; Liu et al., 2023).

The aim of this chapter is to examine how NbS fits into the framework of smart cities. Specifically, it will highlight how these solutions can improve urban resilience, adapt to climate change, and promote sustainability in cities by using the carbon neutrality lens. The objective of this chapter is to present a thorough knowledge of how NbS, by using the technology developments inherent in smart cities, might help alleviate climate-related disasters in urban areas. With the goal of demonstrating how digital technology and natural ecosystems may work together to build more resilient, sustainable, and carbon-neutral metropolitan environments, it critically examines the partnerships between these two types of ecosystems. Moreover, the goal of the chapter is to delineate the obstacles and possibilities related to the integration of NbS in smart cities, providing useful perspectives and policy suggestions for those working in the field of sustainable urban development, including policymakers and urban planners.

4.2 RESULTS AND DISCUSSION

4.2.1 Climate Change, Urbanization, and the Need for Adaptation

i. Overview of Climate Change and Its Impacts
With its profound effects on natural systems, economics, and communities worldwide, climate change stands as one of the most important global concerns of the 21st century. Global temperatures have increased because of human activities like the burning of fossil fuels, deforestation, and industrial operations, which are the main causes of the atmosphere's rising concentration of greenhouse gases. Sea levels are rising, precipitation patterns are changing, heatwaves are occurring more frequently and intensely, and the polar ice caps are melting because of this warming. These alterations have far-reaching effects, upsetting ecosystems, endangering biodiversity, and jeopardizing the supply of basic supplies like food and water. Certain areas, especially low-lying coastal areas, dry zones, and small island states, are more vulnerable to the effects of climate change than others (Wudu et al., 2023; Bhushan & Sharma, 2022). The effects of climate change are most noticeable in metropolitan areas. Cities are particularly vulnerable to disruptions caused by climate change because they are concentrated areas of population, infrastructure, and economic activity. The UHI effect exacerbates heatwaves, which are extremely dangerous for public health, particularly for vulnerable groups including the elderly and people with underlying medical issues. Storm frequency and severity increases have the potential to cause extensive damage to buildings, infrastructure, and basic services while increasing sea levels pose a risk of flooding and erosion to coastal cities. Furthermore, variations in precipitation patterns may result in a shortage of water or, on the other hand, raise the risk of flooding, further taxing metropolitan infrastructure. These effects draw attention to the necessity of developing efficient climate adaptation plans that can strengthen urban resilience and shield natural and human systems from the growing risks associated with climate change (Antal & Bhutani, 2023; Ali et al., 2023).

NbS are essential for lowering the carbon footprint of cities because they improve carbon sequestration and lower energy-related emissions. By using photosynthesis to absorb and store atmospheric CO_2, urban forests, green areas, and wetlands act as natural carbon sinks. For instance, urban trees lessen the environmental impact of emissions from industry and transportation by filtering pollutants and absorbing CO_2. Furthermore, in addition to offering vital ecosystem services like flood mitigation and biodiversity support, wetlands and mangroves contribute significantly to the sequestration of carbon in their biomass and soil. Cities can improve their ability to absorb carbon from the atmosphere and help achieve long-term carbon neutrality goals by incorporating these NbS into urban development (Wang et al., 2024; Zhao et al., 2023; Mandal et al., 2023; Choudhary et al., 2024). NbS lowers energy-related emissions in addition to sequestering carbon by enhancing urban climate regulation and lowering the need for infrastructure that is powered by fossil fuels. By minimizing the demand for heating and air conditioning, green walls and roofs, for example, naturally insulate buildings, lowering emissions from power generation and electricity usage. Likewise, natural stormwater management using permeable pavements and urban wetlands reduces the need for energy-intensive water treatment and pumping systems. Additionally, cool streets and parks reduce urban temperatures, which eases the strain on cooling systems and lowers the need for electricity during peak hours. By implementing NbS in urban planning, communities may create healthier, more climate-resilient ecosystems and drastically reduce emissions, supporting international efforts to become carbon neutral (Bevilacqua, 2021; Gupta & De, 2024; Riffat et al., 2024).

ii. Urbanization Trends and Challenges

Urbanization, which may be described as the growing density of people living in urban areas, is one of the most important demographic developments of the contemporary period. It is anticipated that over 70% of the global population will live in cities by 2050 due to the desire for greater economic prospects, higher quality of life, and better services. However, there are significant obstacles brought about by this quick urbanization. Cities frequently find it difficult to meet the infrastructural needs of expanding populations, which can result in problems like traffic jams, overcrowding, and poor housing. It can also put a burden on public services like healthcare, sanitation, and water supply. Urban area growth sometimes comes at the price of natural ecosystems, resulting in biodiversity loss, environmental deterioration, and a reduction in green spaces which are essential for preserving ecological balance and offering urban dwellers recreational opportunities (Alagirisamy & Ramesh, 2022; Pandey & Ghosh, 2023). Furthermore, socioeconomic disparities that are frequently made worse in urban settings add to the difficulties associated with urbanization. Significant differences in wealth, access to services, and living conditions can be found in many cities, especially in developing nations. Millions of urban inhabitants who live in informal settlements without proper infrastructure or secure housing are especially susceptible to the effects of climate change. These places are prone to climate-related disasters since they are frequently found in high-risk areas like floodplains or unstable hillsides. Integrated planning strategies that not only support population increase but also improve urban resilience, lessen environmental effects, and advance social fairness are needed to address the problems posed by urbanization (Suel et al., 2024; Datola, 2023).

NbS reduce energy demands for cooling by naturally regulating urban temperatures, thereby lowering reliance on air conditioning and decreasing carbon emissions. Green roofs and walls provide insulation, reducing indoor temperatures and minimizing the need for artificial cooling in buildings. Similarly, urban forests and tree-lined streets offer shade, mitigating the UHI effect and reducing overall energy consumption in densely populated areas. Studies have shown that shaded areas can lower surface temperatures by several degrees, significantly cutting electricity demand for cooling systems during peak summer months. Additionally, permeable surfaces and water features, such as ponds and wetlands, help cool the surrounding environment by facilitating evaporative cooling. By decreasing the need for energy-intensive cooling methods, NbS not only enhance urban climate resilience but also contribute directly to carbon neutrality by reducing emissions from power generation (He et al., 2023; Mihalakakou et al., 2023; Tan et al., 2023; Kumar et al., 2024).

iii. Climate-Related Disasters in Urban Areas
Globally, cities are increasingly at risk from climate-related disasters due to extreme weather occurrences that are made worse by climate change. One of the most frequent and damaging climate-related disasters in urban environments is flooding, which frequently results from a mix of high rainfall, insufficient drainage infrastructure, and the impermeable nature of urban surfaces. Floods have the potential to cause major economic losses, displace populations, interfere with transportation networks, and cause considerable damage to infrastructure. Sea level rise and storm surges, which can submerge low-lying communities, damage coastlines, and worsen the effects of tropical cyclones, pose a particular threat to coastal cities. Another significant hazard from climate change is heatwaves, which occur more frequently and with greater severity. The UHI effect is causing extremely high temperatures in cities. These heatwaves can have devastating health impacts, particularly for vulnerable populations, and can strain energy systems as the demand for cooling increases (Clarke et al., 2022; Hernández-Delgado, 2024; Khorat et al., 2024). Other climate-related calamities like droughts, which can cause water shortages, decreased agricultural production, and increased food insecurity, are also becoming more likely to affect urban areas. Cities in areas vulnerable to wildfires are also experiencing longer and more violent fire seasons, endangering people's lives and property as well as the quality of the air. Because urban systems are interrelated, the effects of climate-related disasters can ripple through several sectors, creating more difficult problems. For instance, a flood could cause public health

emergencies in addition to damaging infrastructure and interrupting water and sanitation supplies. The intricacy and gravity of these hazards highlight the necessity of all-encompassing approaches to catastrophe risk reduction that integrate NbS to augment the resilience of cities (D'Evelyn et al., 2022; Mavrouli et al., 2022).

iv. The Urgency for Climate Adaptation in Cities
Urban regions urgently need to adapt to climate change due to the increasing frequency and severity of climate-related disasters. To preserve people's lives, means of subsistence, and vital infrastructure from the effects of climate change, cities must adapt by adopting proactive steps to lessen their vulnerability to these effects. The risks are especially significant since metropolitan regions concentrate on assets and people. The consequences of climate change in cities could be disastrous in terms of the economy, society, and environment if appropriate adaptation measures are not taken. The rising speed of urbanization, which is frequently coupled with insufficient planning and investment in robust infrastructure, intensifies this need even more. Urban growth is surpassing municipal governments' ability to provide basic services, let alone implement full climate adaptation measures, in many cities, especially in developing nations (Abbass et al., 2022; Beitelmal et al., 2024). It is now necessary, not discretionary, to incorporate climate adaptation into urban design and development. To increase their resistance to the effects of climate change, cities need to take a multifaceted approach that incorporates both manmade and NbS. For example, adding green infrastructure—parks, wetlands, and urban forests—can enhance air and water quality, reduce the impact of extreme weather, and have positive social and economic implications. Furthermore, by facilitating data-driven decision-making, real-time monitoring, and more effective resource management, utilizing smart city technologies can increase the efficacy of these solutions (Liu et al., 2023; Olaniyi et al., 2023). Due to the urgency of climate adaptation, governance institutions also need to be rethought. To guarantee that adaptation measures are inclusive, comprehensive, and sustainable, there is a need for increased cooperation between the public and private sectors, as well as civil society.

4.2.2 Nature-Based Solutions: Concept and Relevance

i. Definition and Principles of NbS
The term "NbS" refers to approaches that address societal issues such as catastrophe risk reduction, food security, water management, and climate change by preserving, managing sustainably, and restoring natural or modified ecosystems. The fundamental principle of NbS is to use nature's resources to serve people in many ways at once, such as increasing human well-being, biodiversity, and economic growth. NbS places a strong emphasis on collaborating with nature to create solutions that are resilient, sustainable, and adjustable to shifting environmental conditions, in contrast

to traditional techniques that frequently rely on artificial infrastructure. Examples of NbS include reclaiming wetlands to lower the danger of flooding, establishing urban trees to lessen the impact of heat waves, and building green walls and roofs to control stormwater and enhance urban air quality (Debele et al., 2023; Ferreira et al., 2022). NbS emphasizes the integration of ecological, social, and economic factors and is based on the concept of sustainability. First, to guarantee that the solutions are acceptable and successful, NbS must be context-specific, considering the local environmental, social, and economic constraints. A deep comprehension of the relevant ecosystems and the requirements of the communities that rely on them is necessary for this. Second, NbS are made to offer several advantages, or co-benefits, in various fields and contexts. For instance, a well-planned urban park can concurrently promote mental health, lower heat, boost biodiversity, and raise property prices. Third, NbS stresses the significance of participatory methods and stakeholder involvement. Involving local communities, indigenous peoples, and other relevant parties in the development, execution, and administration of Net-benefit systems (NbS) guarantees that the solutions are egalitarian, inclusive, and sensitive to the needs of the people most impacted by climate change and other issues. Lastly, to ensure their long-term sustainability and efficacy, NbS must be flexible and adaptive, able to change in response to fresh knowledge, shifting circumstances, and new problems (Esraz-Ul-Zannat et al., 2024; Sowińska-Świerkosz & García, 2022; Susca, 2022).

ii. Categories of Nature-Based Solutions

a. Ecosystem-Based Adaptation

One important subcategory of NbS is ecosystem-based adaptation (EbA), which emphasizes using ecosystem services and biodiversity to assist communities in adjusting to the negative consequences of climate change. EbA increases the resilience of human communities and ecosystems by incorporating natural processes into initiatives for climate adaptation. Restoring natural habitats in urban areas, such as wetlands, mangroves, and forests, might contribute to EbA since these areas serve as barriers against climate-related calamities, including heat waves, storms, and floods. These ecosystems offer vital functions like temperature control, carbon sequestration, and water purification in addition to shielding urban areas from severe weather. Urban trees, for example, have the potential to greatly mitigate the UHI effect, lowering temperatures and lowering the energy required for cooling, therefore helping to achieve carbon neutrality (Oktaviani & Masjud, 2024; Paudel et al., 2024). EbA implementation in urban settings necessitates a thorough comprehension of regional ecosystems as well as proactive participation from stakeholders, including nearby residents, municipal planners, and environmental specialists. Effective EbA tactics are tailored to the ecological, social, and economic circumstances of the region and are based on context. They frequently offer several additional benefits, such as boosting biodiversity, strengthening the quality of the air and water, and developing recreational areas that improve human well-being. Additionally, EbA encourages adaptive management techniques that may change as environmental circumstances do, guaranteeing that the solutions continue to work in the face of climate change. Ecosystem services-based architecture (EbA) contributes to wider sustainability objectives like carbon neutrality and biodiversity preservation while also improving urban resilience (Jabbar et al., 2022; Marques et al., 2022; Vikas & Hari, 2023).

b. Green Infrastructure
A network of natural and semi-natural systems that provide a variety of ecosystem services and support urban resilience, sustainability, and carbon neutrality is referred to as "green infrastructure" in urban settings. In contrast to traditional gray infrastructure, which depends on designed fixes, green infrastructure makes use of ecosystem services to solve urban problems. Urban parks, green roofs, street trees, green corridors, and community gardens are a few prominent instances of green infrastructure. These components are essential for controlling stormwater runoff, mitigating the impact of the UHI, boosting air quality, and making cities more livable overall. For example, rainfall can be absorbed by green walls and roofs, lowering runoff and the chance of floods while also insulating buildings and requiring less energy for heating and cooling (Hanna et al., 2024; Milani & Nia, 2024; Almeida et al., 2023). Green infrastructure is an integral part of NbS as it provides multiple co-benefits that align with the goals of carbon neutrality and urban sustainability. Green infrastructure reduces the effects of climate change, sequesters CO_2, and improves biodiversity by integrating vegetation into the urban fabric. Furthermore, urban green areas promote social cohesiveness, enhance mental health, and provide leisure possibilities, all of which enhance the general well-being of the urban population. To guarantee that green infrastructure is implemented in a way that maximizes its ecological and social benefits while integrating seamlessly with current urban systems, rigorous planning and design are necessary. Furthermore, to guarantee the success and sustainability of green infrastructure projects, cooperation between many sectors—such as urban planning, landscape architecture, environmental management, and community organizations—is frequently necessary (Lampinen et al., 2022; Jabbar et al., 2022).

c. Blue Infrastructure
The term "blue infrastructure" refers to the water-related elements of urban ecosystems, such as lakes, rivers, wetlands, canals, and coastal zones, which support urban resilience and climate adaptation while also offering vital ecological services. Like green infrastructure, blue infrastructure is essential for controlling urban water cycles, lowering the risk of flooding, and improving city livability. For example, riparian buffers and wetlands function as natural sponges, soaking up extra precipitation during storms and lessening the chance of flooding in cities. Coastal

blue infrastructure shields coastal cities from the effects of climate change by acting as a natural barrier against storm surges and sea level rise, such as mangroves and coral reefs. Additionally, blue infrastructure promotes biodiversity by giving different aquatic and terrestrial species somewhere to live, enhancing the ecological well-being of urban environments (Battemarco et al., 2022; Ariyarathna et al., 2023; Gies, 2022; Strain et al., 2022). Blue infrastructure is important for sequestering carbon when it comes to carbon neutrality. Among the most effective natural carbon sinks are coastal and marine habitats, which can store significant amounts of carbon over extended periods. Examples of these ecosystems include seagrass beds, mangroves, and salt marshes. Urban planning that incorporates blue infrastructure not only lessens the effects of climate-related disasters but also advances more general sustainability objectives by boosting biodiversity, strengthening water quality, and providing urban dwellers with recreational possibilities. A comprehensive strategy for water management that considers the full hydrological cycle, from rainwater collection and storage to wastewater treatment and release, is necessary for blue infrastructure projects to be effective (Raw et al., 2023; Hamel & Tan, 2022; Pinto et al., 2023). The preservation, restoration, and sustainable management of these natural systems in the face of urban growth and climate change also require cross-sectoral collaboration from water management authorities, urban planners, environmental scientists, and local communities, which is essential to the success of blue infrastructure.

iii. Benefits of NbS in Urban Areas

a. Environmental Benefits

NbS offers substantial environmental benefits in urban areas, playing a crucial role in enhancing biodiversity, improving air and water quality, and mitigating the impacts of climate change (Figure 4.1). The capacity of NbS to preserve and repair natural ecosystems which are frequently severely damaged due to urbanization in urban environments

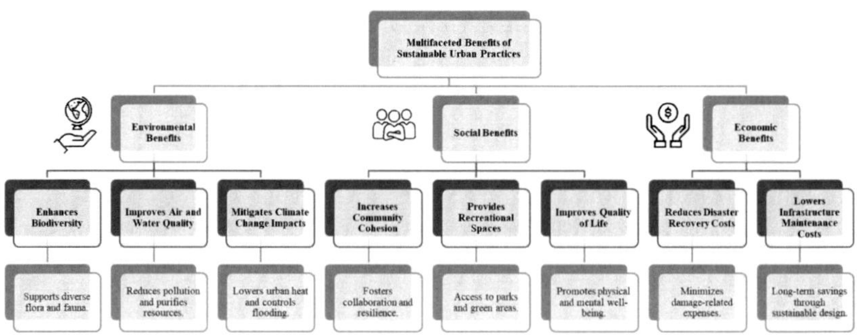

FIGURE 4.1 Overview of environmental, social, and economic benefits of sustainable urban practices.

is one of its most important environmental advantages. For example, the establishment of parks and urban forests, as well as the restoration of wetlands, can greatly increase the biodiversity of the area by serving as habitats for a variety of species, such as pollinators, birds, and other wildlife. In addition to being vital pathways for species migration, these green spaces enhance biological connectedness both inside and between metropolitan regions. Additionally, NbS plays a crucial role in urban efforts to achieve carbon neutrality by simultaneously sequestering CO_2 and improving air quality by absorbing pollutants and particle matter (Kabisch et al., 2022; Dickinson & Ramalho, 2022). Furthermore, NbS plays a crucial role in the sustainable management of urban water resources by reducing the likelihood of climate-related calamities like droughts and floods. Permeable pavements, rain gardens, and green roofs are examples of green infrastructure that improve rainwater infiltration and storage naturally. This lessens the strain on traditional drainage systems and lowers the risk of urban flooding. Restored rivers, lakes, and wetlands are examples of blue infrastructure that help manage floods by serving as organic buffers that soak up extra water during periods of high precipitation. By filtering pollutants and sediments, these water-related NbS also enhance water quality, safeguarding aquatic habitats and guaranteeing a cleaner water supply for urban people. All these environmental advantages of nitrogen-based soil strengthen metropolitan areas' resilience, enabling them to better endure and adjust to the challenges presented by climate change (Baker et al., 2022; Boateng et al., 2023; Ganapathi et al., 2024).

NbS are a crucial tactic for reducing climate change and reaching carbon neutrality because of their high potential for sequestering carbon. By using photosynthesis to absorb and store atmospheric CO_2, urban forests, wetlands, mangroves, and green areas serve as natural carbon sinks. Large-scale afforestation and reforestation projects have the potential to sequester millions of tons of carbon over time, whereas a single mature tree may absorb about 22 kg of CO_2 yearly. Peatlands and wetlands are two of the best habitats for storing carbon because they keep organic matter in wet conditions, which inhibits decomposition and long-term carbon release. Furthermore, carbon is stored in biomass and sediments by blue carbon ecosystems—like mangroves, seagrass beds, and salt marshes—up to ten times faster than by terrestrial forests. In addition to lowering atmospheric carbon levels, NbS improves biodiversity, air and water quality, and climate resilience by protecting and rehabilitating these ecosystems. Long-term carbon reduction initiatives can be greatly aided by incorporating NbS into environmental policies and urban design, which also benefits ecosystems and communities (Seddon et al., 2021; Macreadie et al., 2021; Psistaki et al., 2024; Sudarma et al., 2024).

b. Social and Economic Benefits

Apart from their favorable effects on the environment, NbS offers noteworthy social and economic advantages that bolster the general sustainability and livability of metropolitan regions. Improving public health and well-being is among the greatest direct societal advantages of NbS (Figure 4.1). Parks, urban woods, and community gardens are examples of green spaces that provide people with opportunities for leisure, socializing, and rest—all of which are critical for both physical and mental well-being. These areas can improve overall public health outcomes by lowering stress, promoting physical exercise, and reducing the prevalence of mental health problems. Furthermore, NbS can lower the prevalence of heat-related illnesses, especially in vulnerable groups like the elderly and children, by reducing the effects of UHIs through increased vegetation and shaded spaces (Jones et al., 2022; Bulkeley et al., 2023; Pereira et al., 2024). From an economic standpoint, NbS provides more affordable options than traditional gray infrastructure since they offer ecosystem services that eliminate the need for pricey constructed solutions. For example, by insulating buildings and decreasing the demand for heating and cooling, green walls and roofs can drastically reduce energy expenditures and help achieve carbon neutrality. Economic potential may also arise from the application of NbS, especially in the domains of landscape architecture, urban planning, and environmental management. Green and blue infrastructure design, building, and maintenance jobs, including urban gardening, wetland restoration, and sustainable water management, can boost local economies and create job opportunities. Additionally, by enhancing the visual attractiveness and environmental quality of metropolitan areas, NbS can draw investment and raise property values. The social and economic benefits of NbS, therefore, not only enhance the resilience and sustainability of cities but also contribute to the overall quality of life for urban residents (Carvalho et al., 2022; Perera et al., 2024).

iv. Challenges and Barriers to Implementing NbS

The implementation of NbS in urban areas, particularly through the lens of carbon neutrality, presents a range of challenges and barriers that must be addressed to realize their full potential. The difficulty of incorporating NbS into the current frameworks for urban planning and governance is one of the main obstacles (Figure 4.2). Fragmented governance, in which several departments, agencies, and stakeholders have overlapping responsibilities, is a common feature of urban environments. Implementing NbS that necessitate cross-sectoral collaboration, including green infrastructure projects that include urban planning, water management, and environmental conservation, can be challenging due to coordination issues brought on by this fragmentation. Additionally, it is more difficult for city planners and legislators to prioritize and justify NbS over conventional gray infrastructure due to the lack of standardized metrics and tools to assess the technology's efficacy, particularly in terms

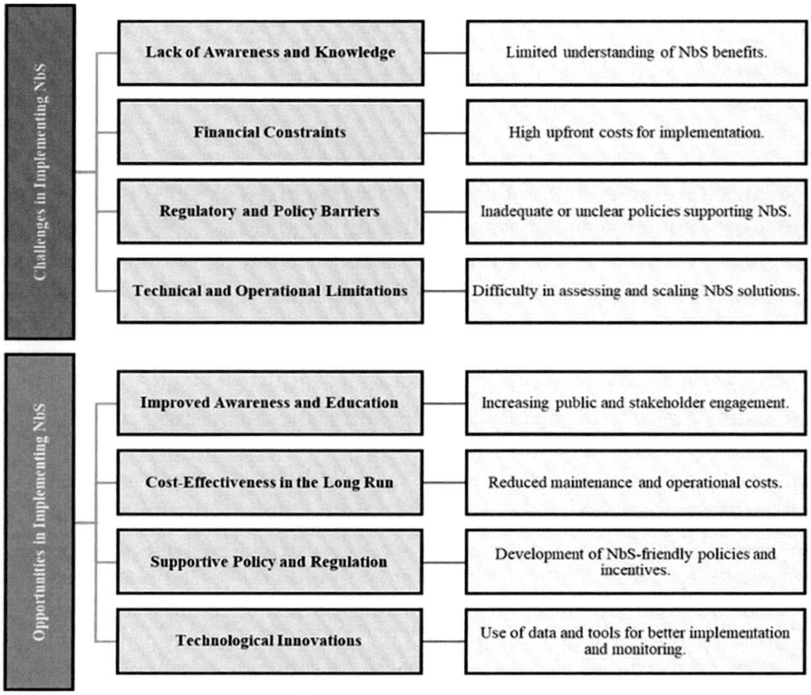

FIGURE 4.2 Challenges and opportunities in implementing NbS, depicting key barriers and potential solutions.

of carbon sequestration and wider sustainability outcomes (Castelo et al., 2023; Dickey et al., 2022; Dorst et al., 2022). The NbS-related budgetary and economic limitations represent another important obstacle. Even though NbS might provide co-benefits and long-term cost advantages, implementing them can require a sizable initial investment, especially in highly populated urban regions where land is expensive and scarce. Furthermore, the advantages of NbS frequently materialize over longer periods of time, which deters investors and decision-makers who could be preoccupied with electoral cycles or short-term profits. This is made even more difficult by the difficulty in valuing and quantifying the numerous co-benefits of NbS, which are frequently overlooked in conventional cost-benefit evaluations and include better public health, higher biodiversity, and stronger social cohesion. Furthermore, there could not be enough public knowledge or comprehension of the advantages of NbS, which could cause stakeholders or communities more accustomed to traditional infrastructure solutions to oppose it. Innovative financing techniques, like public-private partnerships (PPPs), green bonds, and community-based initiatives, as well as increased efforts to increase awareness and develop capacity among urban planners, policymakers, and the public, are needed to overcome these financial, economic, and social barriers (Le Coent et al., 2023; Andrae, 2024; Thompson et al., 2023).

4.2.3 Smart Cities and Climate Change Adaptation

i. Smart City Concept and Components
The Smart City concept represents an innovative approach to urban management, where cutting-edge technologies and data-driven solutions are used to improve metropolitan areas' sustainability, livability, and efficiency. Smart cities optimize the operation of municipal services and infrastructure by utilizing a broad range of digital tools, such as cloud computing, big data analytics, AI, and the IoT. Intelligent transportation systems that lower pollution and traffic, smart water management systems that guarantee effective use of water resources, and smart grids for energy management are important elements of smart cities. Additionally, smart buildings integrate automation and cutting-edge materials to reduce energy consumption and improve occupant comfort. Similarly, smart waste management systems increase recycling rates and decrease landfill usage. These elements are linked by a digital infrastructure that permits data collection, monitoring, and analysis in real-time, empowering city administrators to take well-informed decisions that support resilience and sustainability (Li et al., 2022; Lv & Shang, 2023; Ahmed et al., 2024). The smart city idea places a strong emphasis on participatory governance and citizen participation. To increase accountability, openness, and responsiveness to community demands, smart cities frequently include communication platforms between residents and local government. This comprises online portals and mobile apps that let locals report problems, get services, and take part in decision-making. Furthermore, the creation of sustainable urban environments—which include green areas, infrastructure that is conducive to pedestrian traffic, and energy-efficient structures intended to lower the urban carbon footprint—is given top priority in smart cities. Smart cities seek to build environments that are not only more resilient and efficient but also more egalitarian and inclusive, ensuring that all citizens benefit from the advances in urban living circumstances. They do this by fusing cutting-edge technology with sustainable urban design (Przeybilovicz et al., 2022; Terdiman & Angert, 2023; Umoh et al., 2024).

ii. The Role of Technology in Urban Adaptation
Technology is essential to urban climate change adaptation because it provides resources and solutions that increase cities' resistance to the effects of climate change. Predictive analytics and early warning systems, which use information from sensors, satellites, and weather models to forecast and track extreme weather events like floods, heat waves, and storms, are two important ways that technology helps cities adapt. Cities can reduce potential damage and save lives by utilizing these systems to enable preventive measures, including activating emergency response plans, publicizing warnings, and allocating resources to susceptible locations. Technology also makes it easier to monitor and maintain vital urban infrastructure, such as transport networks, water supply systems, and electricity grids, so that they can resist and quickly recover from disruptions brought on by climate change (Jafari

et al., 2023; Ojo, 2024). Furthermore, technology facilitates the adoption of sustainable urban practices and adaptive infrastructure. For instance, smart networks facilitate the integration of renewable energy sources and more effective energy distribution, which helps cities lower their carbon emissions while preserving a steady supply of electricity under severe weather conditions. Like this, smart water management systems minimize the risk of flooding and guarantee water availability during droughts by managing stormwater, detecting leaks, and optimizing water usage using sensors and automation. Technology-driven green infrastructure, such as automatic irrigation systems on green roofs that improve urban cooling and water management, also helps with urban adaptability. Technology also makes it easier to gather and analyze massive statistics, which enables data-driven decision-making in urban planning and the creation of adaptable solutions that are customized to the unique requirements and vulnerabilities of various metropolitan regions (Agboola & Tunay, 2023; Krishnan et al., 2022; Bibri, 2022).

iii. Integrating NbS in Smart City Planning

a. Digital Tools and NbS

The integration of NbS into smart city planning is becoming more widely acknowledged for improving urban resilience and sustainability. Planning, executing, and monitoring NbS within the context of smart cities are made possible by digital technologies, including Geographic Information Systems (GIS), remote sensing, and simulation models (Petit, 2022). Planners can identify places where NbS can be most effective in minimizing heat islands, managing stormwater, or boosting biodiversity by using GIS, for example, to map and analyze urban green spaces (Cortinovis et al., 2022). For tracking the effectiveness of NbS over time, remote sensing technologies offer real-time data on environmental factors, including vegetation health, water levels, and changes in land usage. By using simulation models, urban planners may forecast the results of various NbS scenarios, such as how urban forests will affect the quality of the local air or how well wetlands will mitigate floods, enabling them to implement the best possible solutions (Kumar et al., 2021). The incorporation of NbS into the current urban infrastructure is also made easier by these digital tools, guaranteeing that old gray infrastructure is enhanced and complemented by green and blue infrastructure. To optimize stormwater retention and lower building energy consumption, digital platforms can be utilized in the design of green roofs. IoT sensors that track biodiversity, air quality, and soil moisture in real-time enable prompt interventions and modifications. Digital tools also help maintain and manage NbS. Cities can maximize the design, implementation, and long-term management of NbS, guaranteeing that it yields the greatest possible environmental, social, and economic advantages, by integrating these technologies into smart city planning (Li et al., 2021a; Baskar et al., 2024).

b. Data-Driven Decision-Making

Data-driven decision-making is a cornerstone of incorporating NbS into the design of smart cities, offering the body of data required to maximize the choice, creation, and administration of these solutions. Cities may use the massive volumes of data produced by smart city technologies—such as social media platforms, environmental sensors, and satellite imagery—to make well-informed decisions about where and how best to deploy NbS. For example, information on UHIs can be used to pinpoint neighborhoods that would most benefit from green roofs or tree planting, while flood risk models can direct the construction of rain gardens or the restoration of wetlands in regions that are vulnerable to flooding. This data-driven strategy guarantees that NbS is in line with more general sustainability and carbon neutrality objectives in addition to being efficient in tackling urban issues (Vögt et al., 2023; Shao & Kim, 2022; Gupta et al., 2024). Additionally, data-driven decision-making facilitates adaptive management and ongoing observation of NbS. City planners can evaluate the success of NbS and make the required modifications to improve its performance by examining real-time data on environmental factors, such as temperature, water levels, and air quality. Data analysis may indicate the need for more irrigation or shade if a green infrastructure project is not providing the anticipated cooling benefits, for instance. To provide a whole picture of NbS's worth to the city, data-driven methods also make it possible to evaluate the co-benefits of the program, such as increases in property values, social cohesion, and public health (Anderson et al., 2023; Mohammed et al., 2023). This comprehensive knowledge enables more effective dissemination of the advantages of NbS to interested parties, including the public, investors, and legislators, promoting their broader acceptance and incorporation into urban planning.

iv. Case Studies of Smart Cities Implementing NbS

a. Global Case Studies

NbS have been effectively incorporated into smart city programs by several cities worldwide, indicating the potential of these methods to improve urban resilience and sustainability. Singapore, also called a "City in a Garden," is one well-known example. To address its urban issues, it has established a vast network of green and blue infrastructure. The conversion of a concrete canal into a naturalized river in Singapore's Bishan-Ang Mo Kio Park is a perfect example of how to create a floodplain that reduces flooding while boosting biodiversity and offering recreational space. The project tracks environmental conditions and water levels using digital monitoring devices, guaranteeing the NbS's efficacy and supplying useful information for upcoming urban planning initiatives. Singapore has improved the quality of life for its citizens, lessened the impact of its UHI, and managed stormwater more skillfully thanks to the integration of NbS with smart technologies (Prothi et al., 2024; Al-Kodmany, 2024; Gopinadh Garre, 2024).

Copenhagen, which has included NbS in its Climate Adaptation Plan, is another example. To absorb rainfall and lower the risk of floods, the city has included pocket parks, permeable pavement, and green roofs in its urban landscape. Data-driven decision-making is the foundation of Copenhagen's strategy, which optimizes the location and architecture of NbS using real-time information from weather stations, IoT sensors, and hydraulic models. Copenhagen has been a trailblazer in urban climate adaptation thanks to its cloudburst management strategy, which integrates NbS with conventional infrastructure. This plan has greatly reduced flood risks and helped the city reach its goal of becoming carbon neutral by 2025. These case studies highlight the potential for cities to address climate concerns while encouraging sustainability and resilience through creative, data-driven initiatives, demonstrating the efficacy of incorporating NbS into smart city planning (Negrello, 2022; Hansson et al., 2023). A city renowned for its creative response to climate change, Rotterdam has made significant use of NbS in its plan to reduce flooding and improve urban sustainability. The city, which is below sea level, has serious problems with rising sea levels and strong rains. Rotterdam has created a thorough water management plan that incorporates blue and green infrastructure into its urban design to solve these problems. The Water Square Benthemplein is a prominent project that serves as a stormwater management system and a public park. During periods of high precipitation, the square is intended to collect and temporarily retain rainwater, lessening the strain on the city's drainage system and averting floods. The region is used as a recreational area when it isn't being used for storing water, which makes the city more livable. Rotterdam also uses real-time data monitoring and digital tools to efficiently manage its water-sensitive urban architecture. The city is equipped with sensors to track water levels, meteorological conditions, and the effectiveness of green infrastructure. City planners can make well-informed decisions on water management and modify NbS operations as necessary thanks to this data-driven approach. By combining NbS with cutting-edge technologies, Rotterdam has reduced energy consumption linked to conventional flood control infrastructure, improving its resilience to climate change and advancing the city's efforts to become carbon neutral (Bona et al., 2022; Tozer et al., 2022b).

Another notable example of a smart city that has effectively incorporated NbS into its urban planning to combat climate change and enhance urban sustainability is Melbourne, Australia. By 2040, the city's Urban Forest Strategy hopes to raise the percentage of tree canopy cover from 22% to 40%. This would improve air quality, make the city more resilient to heatwaves, and aid in the sequestration of CO_2. A strong data-driven foundation underpins the approach, and it makes use of GIS and remote sensing technology to map current green spaces and pinpoint locations that would most benefit from more tree planting. In-depth community involvement is another component of the technique that lets locals choose the kinds of trees to plant and where to grow them. The city of Melbourne's larger objective of becoming carbon neutral is strongly related to its Urban Forest Strategy. The city lowers the energy required for cooling during hot weather by expanding urban vegetation, which also helps to mitigate the UHI effect and sequester CO_2. To ensure that the urban forest survives even in the face of water constraints, Melbourne has also installed smart irrigation systems that optimize water consumption using sensors and weather data. Melbourne is now a role model for other cities looking to improve urban sustainability through NbS because of its combination of NbS with smart technologies and community involvement (Barker et al., 2024; Sousa-Silva et al., 2023; Parihar & Birman, 2024; Adams et al., 2023).

To manage stormwater, lower pollution, and increase urban resilience, New York City (NYC) has put into practice a comprehensive green infrastructure plan that incorporates NbS. The concept is to use permeable pavements, rain gardens, bioswales, and green roofs to collect rainfall and lessen the load on the city's deteriorating sewer system. The city's green infrastructure components are positioned with purpose, especially in regions that are vulnerable to floods and CSOs, which happen when excessive rainfall overwhelms the sewer system and releases raw sewage into bodies of water. NYC's Green Infrastructure Plan is based on a data-driven methodology, evaluating NbS performance and stormwater management effects through intensive modeling and monitoring. Additionally, the city has launched a Green Infrastructure Grant Program to encourage private property owners to use NbS, thus increasing the program's impact and reach. NYC has made great strides in strengthening its resilience to climate change, enhancing water quality, and advancing toward its carbon neutrality goals by lowering energy use associated with conventional stormwater management techniques by incorporating NbS into its urban planning and utilizing smart technologies (Balk et al., 2024; Radfar et al., 2024; Foster et al., 2024).

Barcelona's "Superblocks" concept, which reorganizes city streets to prioritize pedestrians, bicycles, and green spaces over automotive traffic, is an innovative approach to urban sustainability. This program is a component of Barcelona's larger plan to lower air pollution, lessen the impact of the UHI, and improve living conditions for locals. Every Superblock consists of nine city blocks with traffic limited to the outside, making the interior streets green, pedestrian-only public areas. These areas frequently have playgrounds, urban gardens, and walks surrounded by trees. By lowering temperatures, boosting biodiversity, and improving air quality, these features help the city adapt to climate change. Digital tools that track traffic patterns, air quality, and green space health complement Barcelona's strategy. To make sure that the Superblocks program successfully achieves its social and environmental objectives, the city uses this data to extend and improve it. Barcelona has improved the sustainability, resilience, and livability of its urban environment by incorporating NbS into its urban design and utilizing technology for management and monitoring. This has helped the city achieve its long-term objectives of lowering carbon emissions and enhancing urban sustainability (Anguelovski et al., 2022, 2023; Macaione et al., 2024). One of the fastest-growing cities in China, Shenzhen, has adopted the "Sponge City" initiative, a national program that uses NbS to improve urban water management. The Sponge City concept lowers flood risks and improves water quality by absorbing and filtering rainwater through permeable surfaces, green roofs, wetlands, and rain gardens. By integrating these NbS across its urban environment, Shenzhen has built a network of blue and green infrastructure that resembles natural hydrological processes. This strategy helps the city become more resilient to climate change by reducing pollutants, mitigating flooding, and replenishing groundwater resources. Cutting-edge technology like IoT sensors and GIS mapping, which track NbS performance in real-time and supply data for adaptive management, are supporting Shenzhen's Sponge City program. The city's dedication to incorporating NbS into urban design is consistent with its overarching objectives for sustainability, which encompass carbon neutrality. Shenzhen's Sponge City initiative, which aims to combine development with environmental sustainability, is a model for other rapidly urbanizing locations. It does this by lowering the need for energy-intensive water management systems and improving the ability of the urban environment to sequester carbon (Li et al., 2023; Han et al., 2023; Song, 2022; Jiang et al., 2022).

b. Lessons Learned from Case Studies

In addition to regional projects in Asia, Europe, and North America, the case studies of Singapore, Copenhagen, and New York City offer important insights into how to apply NbS in urban contexts successfully. The significance of incorporating NbS into more comprehensive urban planning and policy frameworks is one important lesson. Successful NbS implementations, like those in Singapore and Copenhagen, have been achieved by integrating these solutions into comprehensive urban development plans, which guarantee that NbS are not viewed as standalone initiatives but rather as essential parts of urban infrastructure. Since NbS is in line with the city's economic, social, and environmental objectives, this strategy guarantees long-term sustainability and scalability. These case studies also demonstrate the importance of robust governance frameworks and multi-stakeholder cooperation, as demonstrated by Copenhagen's Cloudburst Management Plan, which entails cooperation between local communities, private sector partners, and government agencies (Hansson et al., 2023; Prothi et al., 2024; Balk et al., 2024; Parihar & Birman, 2024). One more important takeaway from these case studies is how innovation and technology can improve NbS's efficacy and scalability. Smart sensors, data analytics, and the IoT have been used by cities like Copenhagen and Singapore to optimize NbS design, deployment, and management, making it more flexible and responsive to changing circumstances. These case studies also highlight how crucial public involvement and community engagement are to the accomplishment of NbS projects. As evidenced by New York City's "Rebuild by Design" effort, including local communities in the planning, construction, and upkeep of NbS promotes a sense of ownership and guarantees that NbS satisfies the requirements of locals. In the end, these experiences show how NbS has the power to make urban settings more livable, sustainable, and resilient—that is, if they are backed by robust laws, cutting-edge technology, and engaged communities (Li et al., 2023; Liu et al., 2023; Han et al., 2023; Macaione et al., 2024).

4.2.4 Urban Resilience and Nature-Based Solutions

i. Understanding Urban Resilience

a. Definition and Frameworks

The phrase "urban resilience" describes a city's capacity to take in, adjust, and bounce back from a variety of shocks and strains, especially those associated with climate change, like extreme weather, rising sea levels, and long-term environmental degradation. It includes a city's social, economic, and institutional structures in addition to its physical infrastructure. The knowledge that cities are dynamic, complex systems that need to be adaptable and flexible in the face of uncertainty is the foundation of the idea of urban resilience. Numerous frameworks have been created to aid in the evaluation and improvement of urban resilience. One such framework is the City Resilience Framework from the Rockefeller Foundation, which highlights the need to combine the physical, social, and economic elements. This paradigm emphasizes important elements that contribute to a city's overall resilience, such as the diversity of economic resources, the strength of the infrastructure, and the inclusivity of governing systems (Xie et al., 2023; Amirzadeh et al., 2022; Kumar & Mehany, 2022; Silva Martins da Cunha Marinho et al., 2024). The UN-Habitat City Resilience Profiling Tool is another well-known framework that offers a thorough method of evaluating and planning for resilience by emphasizing the systemic links found in urban environments. It pushes cities to create multi-sectoral plans that include sustainable development, climate adaptation, and disaster risk reduction. These frameworks assist cities in recognizing their strengths and weaknesses, allowing them to create focused plans that improve their ability to endure disturbances and bounce back from them. Through the implementation of frameworks of this kind, cities can systematically enhance their resilience, making them more capable of managing acute shocks such as floods and storms, as well as chronic stresses like poverty and environmental degradation (Dianat et al., 2022; Eltinay & Egbu, 2024).

b. Indicators of Urban Resilience

Metrics called indicators of urban resilience are employed to evaluate a city's ability to endure and bounce back from different types of disruptions. These metrics are essential for tracking development and pinpointing areas where resilience planning needs to be strengthened. The resilience and redundancy of vital infrastructure, such as the water, power, and transportation networks, are common indicators. To maintain operations in the event of a catastrophe, a resilient city should, for instance, have many separate and independent sources of water and energy. Furthermore, social resilience indicators—like community cohesion and the inclusivity of governance—are essential for evaluating how well various population segments can react to and recover from disasters. The degree of trust among community members and their involvement in decision-making processes, for example, are indicators of social cohesiveness and have a direct bearing on the community's capacity to assist one another in times of need (Serdar et al., 2022; Osei-Kyei et al., 2023). Indicators of economic resilience, such as the local economy's flexibility and diversity, are equally important. An economy that is resilient is not unduly reliant on any one industry or sector, which enables it to adjust to external factors like changes in the global economy or the effects of climate change. Environmental metrics, such as the condition of natural ecosystems and the presence of green areas, are equally significant since they help the city lessen the effects of climate change and adjust to them. These metrics not only offer a quick assessment of a city's resilience right now, but they also serve as a roadmap for the creation of actions and policies that will gradually increase urban resilience. Cities may track their progress and make well-informed decisions to fortify their resilience against future crises by routinely evaluating these metrics (Thomas, 2023; Epelde et al., 2022).

ii. NbS as a Tool for Enhancing Urban Resilience

Urban resilience can be improved by the application of NbS, especially considering climate change. Natural block structures NbS harness the power of natural ecosystems to offer a multitude of benefits, including temperature regulation, biodiversity enhancement, and flood protection, all of which contribute to the overall resilience of a city. The UHI effect can be considerably reduced, for instance, by green roofs and urban forests. This lowers temperatures and lowers the energy required for cooling, which is especially crucial during heatwaves. In a similar vein, restored wetlands and riverbanks can serve as organic flood defenses by soaking up extra water and lowering the chance that flooding will harm urban infrastructure. By enhancing social cohesion, creating recreational areas, and improving resident quality of life, these natural solutions not only improve social resilience but also increase the resilience of physical infrastructure (Paxton et al., 2023; Shao & Kim, 2022; Clarke et al., 2023). By strengthening cities' ability to adapt, NbS also helps urban resilience. Non-biological systems are self-sustaining and dynamic, adapting to changing environmental conditions over time, unlike traditional gray infrastructure, which frequently requires costly and resource-intensive upkeep. To provide long-term protection against storm surges and coastal erosion, mangroves and other coastal ecosystems, for example, can grow and extend in response to increasing sea levels. NbS also provides flexibility and scalability, enabling cities to adopt solutions that are customized for their unique circumstances and may be expanded or modified as necessary. Cities can build more adaptable, sustainable, and resilient settings that are better suited to address the uncertainties and challenges presented by climate change by incorporating NbS into urban planning and construction (Zhu et al., 2024; Sunkur et al., 2023; Castelo et al., 2023).

iii. Assessing the Impact of NbS on Urban Resilience
The process of determining how well NbS improves a city's ability to endure and bounce back from disruptions connected to climate change and other events is known as the "impact assessment." Cost-benefit analysis, community comment collection, and environmental indicator monitoring are just a few of the quantitative and qualitative techniques commonly used in this assessment. For example, surface runoff reduction and the frequency and intensity of urban flooding occurrences are two indicators of how effective green infrastructure is, such as bioswales or permeable pavements. The evaluation of urban green spaces' effects on public health and well-being can also be done using surveys and health data. For example, one can look at how much more access to nature reduces the risk of heat-related illnesses or enhances mental health (Debele et al., 2023; Nita et al., 2022; Khodadad et al., 2023). Furthermore, the benefits of NbS for resilience can be evaluated by analyzing their role in achieving climate adaptation and carbon sequestration objectives, such as lowering greenhouse gas emissions or improving biodiversity. For instance, urban woods enhance the city's overall ecological resilience by acting as a habitat for wildlife in addition to sequestering carbon. Because the advantages of NbS frequently materialize over time, assessing these effects necessitates long-term observation and data collection. These changes are often monitored and a thorough understanding of how NbS supports urban resilience is provided by using tools such as GIS mapping, remote sensing, and environmental modeling. Cities can fine-tune and maximize the benefits of their NbS plans in terms of resilience, sustainability, and carbon neutrality by using this data-driven approach (Pandey & Ghosh, 2023; Wild et al., 2024).

By increasing efficiency, scalability, and real-time monitoring, adaptive management in conjunction with digital technology improves the effectiveness of NbS in lowering overall carbon footprints. Urban forests, wetlands, and green infrastructure are examples of NbS that naturally store carbon and control the temperature. However, when combined with smart technology, their impact can be greatly increased, allowing for data-driven decision-making for optimal effectiveness. The performance and sustainability of NbS are enhanced by digital techniques, including automation, predictive modeling, remote sensing, AI, and big data analytics, which offer accurate monitoring and predictive modelling. Cities and businesses can improve carbon sequestration, maximize resource utilization, and lessen dependency on energy-intensive infrastructure by incorporating these technologies into NbS projects, which will ultimately reduce overall carbon emissions (Bressane et al., 2024; Raji et al., 2024; Granata & Di Nunno, 2025). Smart urban forestry management, which uses AI-driven sensors and satellite imaging to monitor tree growth, health, and carbon sequestration rates, is one method digital technology maximizes NbS. Based on factors including drought tolerance, carbon absorption capacity, and local biodiversity requirements, machine learning algorithms evaluate climate and environmental data to suggest the ideal tree species for urban reforestation. Furthermore, IoT-enabled irrigation systems minimize water waste by using real-time weather and soil moisture data, which lowers the energy required for water distribution and pumping. This data-driven strategy ensures urban trees' long-term resilience against the effects of climate change while optimizing their effectiveness as carbon sinks (Bressane et al., 2024; Nsoh et al., 2024; Yadav et al., 2024).

Intelligent water management systems are another area where NbS is improved by digital optimization. By dynamically adapting to rainfall patterns, smart stormwater infrastructure—such as sensor-equipped permeable pavements and AI-controlled retention ponds—reduces the need for energy-intensive drainage and wastewater treatment. City planners can predict how NbS, such as wetlands or bioswales, can replace conventional gray infrastructure, drastically lowering emissions and energy use, using digital twins, which are virtual representations of water ecosystems. For instance, Singapore optimizes desalination and water recycling while lowering carbon footprints related to water supply by using AI-driven water management to control its vast green-blue infrastructure (Abedi et al., 2023; Bressane et al., 2024; Chapagain et al., 2024; Sharifi et al., 2024). Additionally, drones and the IoT make it possible to monitor green infrastructure and precision agriculture, which are crucial for lowering carbon emissions. Vertical gardens and smart green roofs with environmental sensors modify fertilizer supply and irrigation according to weather, reducing waste and increasing plant health and carbon sequestration effectiveness. Wetland monitoring systems with AI capabilities can identify shifts in carbon storage capacity, maximizing conservation efforts for CO_2 absorption. Similarly, digital mapping tools ensure that NbS are deployed in the most efficient sites by assisting in the identification of degraded areas appropriate for ecosystem restoration (Alsamhi et al., 2021; Ali et al., 2025). Adaptive management increases productivity, lowers resource usage, and increases carbon sequestration potential by combining digital technologies with NbS. These innovations support a more data-driven, responsive approach to sustainability, ensuring that nature-based interventions achieve their maximum potential in reducing carbon footprints. As urban areas and industries adopt smart NbS strategies, they can accelerate the transition to carbon neutrality while promoting climate resilience, biodiversity conservation, and sustainable development (Bressane et al., 2024; Ali et al., 2025).

iv. Challenges in Enhancing Urban Resilience through NbS
Increased urban resilience through NbS faces considerable obstacles despite its potential. A key obstacle is the incorporation of NbS into current urban planning frameworks, which are frequently dominated by conventional gray infrastructure methodologies. For developers, legislators, and urban planners who might be more used to traditional engineering solutions, this integration necessitates a mentality change. NbS also frequently needs larger spaces than traditional infrastructure, which can be challenging to provide in crowded cities with limited and expensive land. The scope and efficacy of NbS may be restricted by this spatial limitation, especially in urban areas where swift urbanization and population expansion are putting a strain on usable land (Castelo et al., 2023; Petit, 2022). The requirement for sustained NbS maintenance and dedication presents another difficulty. In contrast to gray infrastructure, which is usually built at one time and then requires continuing maintenance, NbS needs continuous management to guarantee its continued efficacy. For instance, wetlands and rivers may require periodic restoration to preserve their ecological services, whereas urban green areas require regular maintenance such as watering, pruning, and pest control. Cities with limited financial and human resources may find it difficult to do this continuing maintenance because it can be resource-intensive and require specialized knowledge and skills. Additionally, compared to traditional infrastructure, the benefits of NbS are sometimes spread across longer durations and may not be immediately visible, making it more difficult to win finance and political support. Innovative funding methods, public-private collaborations, and capacity-building initiatives are needed to overcome these obstacles and guarantee that cities can successfully integrate and maintain NbS into their urban resilience plans (Ferreira et al., 2022; Castelo et al., 2023).

4.2.5 Climate-Related Disasters and NbS in Urban Areas

i. Types of Climate-Related Disasters Affecting Cities
The frequency and intensity of climate-related disasters are rising, with a major impact on metropolitan centers that house a significant portion of the world's population. The primary climate change-related natural disasters that affect cities include heat waves, floods, storms and hurricanes, and droughts.

Flooding is one of the most common and destructive climate-related disasters that affect cities globally. Storm surges, severe rain, and river overflow can all result in flooding. Urban areas are particularly susceptible to floods because of the prevalence of impermeable surfaces like concrete and asphalt, which reduce natural drainage and increase surface runoff. Flooding can cause fatalities, interrupt infrastructure, destroy essential services, and cause financial damages. Cities with poor drainage systems or those in low-lying areas are especially vulnerable. Flash floods have the potential to overrun drainage systems, causing waterlogging, contaminated drinking water sources, and property damage. Climate change exacerbates flooding by raising sea levels, increasing the frequency and intensity of rainfall, and intensifying storms. The danger of flooding increases as cities grow due to the loss of natural vegetation for urban expansion (Kumar, 2021; Jain, 2022). Extended periods of uncomfortable heat that are usually accompanied by excessive humidity are known as heatwaves. Metropolitan areas are more prone to heatwaves because of the UHI effect, which happens when cities experience higher temperatures than surrounding rural regions due to human activity and infrastructure. Heatwaves pose a major threat to public health, particularly for the elderly, young people, and those with underlying medical conditions. They can exacerbate heart and lung ailments, result in heatstroke, and dehydrate people. Heatwaves increase the need for cooling, which raises the possibility of power outages, further taxing the energy system. The frequency, duration, and intensity of heatwaves are all rising due to climate change. Urbanization makes cities more susceptible to extreme heat events because of their dense population and lack of green space, which exacerbates the impact of UHI (Marandi & Main, 2021; Li et al., 2021b).

Typhoons and hurricanes are examples of storms that can produce storm surges, powerful winds, and heavy rainfall. Cities, particularly those close to the coast, could sustain significant damage from these events. Storms and hurricanes have the potential to seriously harm power lines, structures, and other infrastructure. Cities along the shore are particularly vulnerable to storm surges, which raise the risk of floods, erosion, and mortality. Long-term disruptions to water supplies, transit networks, and electrical networks are frequent in the aftermath, significantly impacting daily life and economic activities. Because of the increasing energy that hurricanes and storms receive from warmer sea temperatures, climate change is making them stronger. As sea levels rise, coastal cities are increasingly susceptible to storm surges (Kumar, 2021; Marandi & Main, 2021). Water shortages result from extended periods of unusually low precipitation, or droughts. Urban areas are more susceptible to droughts because of the increased demand for water brought on by expanding populations, especially in dry or semi-arid areas. Water shortage brought on by droughts can have an impact on business, agriculture, and drinking water sources. They may also make urban heatwaves worse, raise the possibility of wildfires, and cause financial losses for industries that depend on water supplies. Water rationing and limits might be required, which would lower living standards. The frequency and intensity of droughts are predicted to rise due to climate change, especially in areas where water scarcity is already a problem. Urban regions need to improve water management techniques, create infrastructure resistant to drought, and increase water conservation to adapt to these changes (Kumar, 2021; Marandi & Main, 2021).

Compared to traditional disaster mitigation strategies, NbS in urban areas have improved sustainability and drastically reduced carbon emissions. Urban green spaces can reduce building energy consumption by up to 25%, which lowers associated CO_2 emissions (Susca et al., 2022). In contrast to traditional flood defenses like seawalls, which have no carbon sequestration benefits, coastal wetland restoration, a crucial NbS method, minimizes storm surge damage while sequestering up to 450 million metric tons of CO_2 per year (Hull, 2024). Furthermore, while traditional infrastructure like concrete flood barriers contributes to embodied carbon emissions rather than reducing them, urban tree planting programs can offset up to 18.9 kg of CO_2 per tree annually (Xiao et al., 2023).

ii. The Role of NbS in Mitigating Climate-Related Disasters

NbS are approaches that leverage ecosystems and natural processes to address societal problems including climate-related disasters. NbS are increasingly recognized as practical and effective strategies to mitigate the impact of climate-related disasters on urban areas. This will be condensed into four categories: drought resilience, storm and hurricane protection, heatwave mitigation, and flood management (Debele et al., 2023). Restoring wetlands, growing green roofs, and designing permeable pavements are examples of NbS for flood control. By improving natural water absorption and retention, these methods lower the chance of flooding. By absorbing surplus rainfall and lowering surface runoff, restoring wetlands in and around cities can serve as natural flood protection barriers. Additionally, wetlands improve water quality by filtering pollutants. Permeable pavements and green roofs let water seep into the ground instead of running off, which reduces the amount of water that enters drainage systems after heavy rains (Ferreira et al., 2021).

Urban green spaces are essential for reducing the effects of heatwaves. Examples of these areas include parks, green roofs, and streets lined with trees. By producing shade, encouraging air circulation, and lowering air temperature through evapotranspiration, these NbS contribute to lessening the impact of the UHI. Urban forests and tree planting are two effective ways to drastically lower city temperatures. In addition to providing shade, trees enhance air quality and lessen the demand for air conditioning. Green walls and roofs insulate buildings, lowering the demand for artificial cooling systems and averting heatwaves. They also enhance urban aesthetics and support biodiversity (Wong et al., 2021). Mangroves, coral reefs, and dunes are examples of coastal ecosystems that can act as organic storm barriers and hurricane breakers. The force of storm surges and waves is absorbed by these NbS, lessening their effect on cities. Mangroves act as a buffer against storm surges, stabilize coasts, and lessen erosion. They also promote carbon sequestration and biodiversity. This entails preserving natural coastal processes while stabilizing beaches and lessening the effects of waves and storm surges using natural materials like plants, sand, and rocks (Jordan & Fröhle, 2022). Restoring natural water cycles, encouraging sustainable farming practices, and using rainwater harvesting are some NbS for drought resistance. During droughts, gathering and storing rainwater can ease the strain on the urban water supply. The pressure on potable water supplies can be decreased by using this water for agriculture. Water retention in the landscape is ensured by restoring and maintaining healthy watersheds, which also enable groundwater recharge and sustain base flows in rivers during dry spells (Kumar et al., 2021).

iii. Case Studies of NbS in Disaster Risk Reduction
Particularly in urban settings, NbS are becoming more widely acknowledged as useful instruments for disaster risk reduction (DRR) and climate adaptation. Let's dissect these case studies and discuss how they are used in urban contexts and incorporated into catastrophe risk management programs.
 a. Restoration of Mangroves for Coastal Defense
 The rehabilitation of mangrove forests along coastlines, especially in Southeast Asia, is one noteworthy case study. Mangroves serve as organic defenses against erosion along the shore and storm surges and tsunamis. Large-scale mangrove restoration, for example, has greatly lessened the impact of typhoons in Vietnam's Mekong Delta, where people report less loss of life and damage to infrastructure. Mangroves provide immediate hazard protection, but they also improve fisheries stocks and provide wood resources, supporting local livelihoods and linking disaster risk reduction with socioeconomic advantages (Oanh et al., 2020; Nyangoko et al., 2022).
 b. Europe's floodplain restoration
 One important NbS tactic to reduce flood hazards in Central and Western Europe is floodplain restoration. An excellent example is the Dutch "Room for the River" initiative, which returned riverbank regions to their natural state so that rivers might overflow during high water events without endangering nearby villages. This strategy lowers the risk of flooding while simultaneously improving the quality of the water, enhancing biodiversity, and creating recreational spaces (Boulton et al., 2013; Cui et al., 2021).
 c. The Management of Forests in the Himalayas Sustainable Forest management techniques have been instituted in the Himalayan region to stabilize slopes and lessen soil erosion because of the regular occurrence of landslides and flash floods. Deep-rooted plant species have been cultivated in conjunction with community-led forestry initiatives to help bind the soil and lessen the risk of landslides. Additionally, due to these efforts, carbon sequestration has risen, advancing more general climate mitigation objectives (Tomar et al., 2023; Verma et al., 2021).

iv. Integrating NbS into Disaster Risk Management Plans
 a. Embedding NbS into Local and National Policies
 A multifaceted strategy that incorporates policy reform, community participation, and cross-sectoral collaboration is needed to integrate NbS into disaster risk management (DRM) planning. Nations with a high risk of climate-related disasters, such as Bangladesh and the Philippines, have started integrating NbS into their national DRM frameworks. To ensure that NbS are considered when allocating resources for disaster preparedness and recovery, they must be synchronized with currently

available disaster risk assessment methods, integrated into land-use planning, and evaluated (Bueb et al., 2021).

b. Funding and Rewarding NbS

One major obstacle to the deployment of NbS is funding. Innovative financial tools are being investigated as a solution, such as insurance programs that incentivize ecosystem restoration and green bonds. Additionally, proving the long-term cost-effectiveness of these solutions in comparison to conventional engineering approaches is also necessary for integrating NbS into catastrophe risk management plans. For instance, research has demonstrated that restoring wetlands can result in lower catastrophe recovery costs and a reduction in the need for costly flood defenses (Beck et al., 2022; Ranjan, 2024).

c. Developing Capabilities and Exchanging Knowledge

At different levels of governance, capacity building is also necessary for the effective integration of NbS into DRM programs. This includes developing venues for knowledge sharing and educating communities, planners, and local government officials on the creation and application of NbS. These exchanges and the provision of technical assistance to nations are greatly aided by international organizations like the United Nations Office for Disaster Risk Reduction (UNDRR) and the International Union for Conservation of Nature (Murthy et al., 2024).

v. NbS in Urban Areas

a. Green infrastructure in urban areas

With the growth of green infrastructure, the integration of NbS in urban areas is becoming more and more apparent. Permeable pavements, green roofs, and urban parks are a few examples of NbS that control stormwater runoff, lessen the impact of the UHI, and enhance air quality. Large-scale green infrastructure initiatives have been put in place by cities like Copenhagen, Singapore, and New York as a component of their climate resilience plans. By creating recreational areas and cutting down on pollution, these programs improve the quality of life in addition to lowering the danger of natural disasters like flooding and intense heat (Biswal et al., 2022; Pauleit et al., 2020).

c. Restoring Rivers in Urban Areas

River restoration, in which towns revitalize deteriorated streams to improve their inherent flood control capabilities, is another urban NbS case study. One well-known example is the restoration of Seoul, South Korea's Cheonggyecheon Stream. The city decreased the risk of flooding, enhanced the quality of the air, and created a lively public place that has rejuvenated the neighborhood by removing a concrete flyover and restoring the stream. This study demonstrates how NbS may complement urban revitalization initiatives, enhancing both urban livability and disaster risk reduction (Han & Kuhlicke, 2021).

c. Community-Based Approaches Strong community involvement is generally required for urban NbS. Planning and maintaining green spaces are often the responsibility of the local community, which helps guarantee that these areas are managed sustainably and are created to suit local needs. In Medellín, Colombia, for instance, the development of green corridors connecting parks and natural areas has been a community-driven initiative. The multifunctional advantages of NbS in urban settings are demonstrated by these corridors, which aid in stormwater management, lower the risk of landslides, and provide cooling effects in highly populated neighborhoods (Kiss et al., 2022; Croeser et al., 2024).

NbS are turning out to be extremely useful instruments for lowering the risk of disaster, particularly considering urbanization and climate change. Through the utilization of ecosystems' innate capabilities, NbS offers an economical, environmentally friendly, and socially inclusive method of catastrophe risk reduction. But to reach their full potential, NbS must be incorporated into disaster risk management strategies at all governmental levels and backed by sufficient funding, legislation, and community involvement. NbS provides a method to improve the quality of urban living while also fostering resilience in urban settings, where the effects of climate change are most noticeable (Han & Kuhlicke, 2021; Biswal et al., 2022; Croeser et al., 2024).

4.2.6 Urban Sustainability through the Lens of Carbon Neutrality

i. Definition and Importance of Urban Sustainability
Urban sustainability is used to describe a city's capacity to function in a way that satisfies current demands without endangering the capacity of future generations to satiate their needs. This idea includes striking a balance between social justice, economic growth, and environmental sustainability. Metropolitan sustainability is a crucial objective because of the growing pressure on resources, infrastructure, and ecosystems caused by the world's population converging in metropolitan areas. By reducing trash, using renewable resources more often, conserving energy and water, and preserving natural habitats, sustainable cities seek to lessen their negative effects on the environment. In addition, sustainable urban development promotes social inclusivity and economic opportunity while working to improve the quality of life for all citizens by guaranteeing access to clean water, air, and green areas (Bibri et al., 2020; Hariram et al., 2023). The reality that cities both significantly contribute to and play a major role in driving solutions for global environmental concerns highlights the significance of urban sustainability. Global resource use, waste generation, and carbon emissions are mostly attributable to urban regions. Therefore, combating climate change and preserving natural ecosystems depend heavily on attaining urban sustainability. Furthermore, the effects of natural disasters, economic downturns, and social injustices are less likely to affect sustainable cities. Afforestation, reforestation, and wetland restoration are examples of NbS for carbon sequestration that can absorb and store large amounts of CO_2; forests alone sequester about 7.6 billion metric tons of CO_2 each year (Sonkar, 2024). Compared to most terrestrial forests, mangrove ecosystems store up

to 1,023 metric tons of carbon per hectare, even though they make up less than 1% of the Earth's surface (Xiao et al., 2024). Furthermore, better farming techniques like agroforestry and cover crops can increase soil carbon absorption by 0.3–1.2 metric tons of CO_2 per hectare annually, greatly aiding in attempts to mitigate the effects of climate change worldwide (Pant et al., 2023). Cities can safeguard vulnerable populations, lessen the negative consequences of growing urbanization, and guarantee the prosperity and well-being of their citizens in the long run by implementing sustainable policies. In addition to being a means of preserving the environment, urban sustainability is an essential framework for developing livable, just, and economically viable communities (Bibri et al., 2020; Naya et al., 2023).

ii. Carbon Neutrality: Concepts and Pathways

The balance between the quantity of CO_2 released into the atmosphere and the amount removed or offset by various methods is referred to as carbon neutrality, sometimes known as net-zero carbon emissions. Reducing greenhouse gas (GHG) emissions to as near to zero as feasible and offsetting any leftover emissions with carbon-absorbing or carbon-capturing activities like carbon capture and storage or reforestation are the steps involved in achieving carbon neutrality. Since it is consistent with the objectives of the Paris Agreement, which aims to keep the rise in global temperatures to well below 2°C over pre-industrial levels with aspirations to restrict it to 1.5°C, the idea of carbon neutrality is essential to efforts being made to tackle climate change worldwide (Fankhauser et al., 2022; Black et al., 2021). Through carbon sequestration, NbS help achieve carbon neutrality. Global reforestation initiatives might sequester up to 205 gigatons of CO_2 over the course of the next century (Bastin et al., 2019; Leltz, 2022). Despite comprising up only 3% of the planet's surface area, peatland restoration is a very effective long-term carbon sink, able to absorb carbon at rates of 0.37 to 1.23 metric tons of CO_2 per hectare per year (He & Roulet, 2023). Furthermore, the combined yearly sequestration of 235 to 450 million metric tons of CO_2 by coastal blue carbon ecosystems—which include mangroves, salt marshes, and seagrasses—contributes significantly to the global carbon neutrality targets (Hull, 2024). Reducing emissions at the source and improving carbon sequestration are two strategies for achieving carbon neutrality. Cities can work toward becoming carbon neutral by pursuing several measures, such as switching to renewable energy sources, improving building and transportation energy efficiency, encouraging sustainable land use, and putting circular economy policies into place that reduce waste. Cities can also make investments in natural climate solutions, such as preserving and reforesting wetlands, forests, and other carbon sinks that are essential for removing CO_2 from the atmosphere. The path to carbon neutrality is arduous and necessitates concerted efforts by the government, business community, and civil society to create and execute low-carbon future-supporting policies, technologies, and behaviors (Möslinger et al., 2023; Seddon et al., 2021).

iii. The Role of NbS in Achieving Carbon-Neutral Cities
Urban forests, which store over 25.6 million metric tons of CO_2 yearly in the U.S. alone, can be included into NbS to help create carbon-neutral cities (Bherwani et al., 2024). Vegetated roofs can absorb up to 375 grams of CO_2 per square meter annually, and urban wetlands and green roofs improve carbon sequestration (Halim et al., 2022). Furthermore, city emissions can be offset by up to 18.9 kg of CO_2 per tree annually through urban tree planting programs, which supports municipal climate action plans for carbon neutrality (Xiao et al., 2023). With the ability to use ecosystems' natural processes to absorb and store carbon, NbS are essential to achieving carbon neutrality in urban areas. They also offer several co-benefits, including increased resilience to climate-related disasters, improved biodiversity, and improved air and water quality. By taking in CO_2 from the atmosphere and storing it in biomass and soils, NbS-containing environments like wetlands, urban forests, and green roofs help with carbon sequestration. Furthermore, by lowering UHI effects, these natural systems aid in controlling regional climates. This can lower the need for energy-intensive cooling, which cuts carbon emissions indirectly (Quagliolo, 2023; Huynh et al., 2024). NbS fosters carbon neutrality by encouraging sustainable urban development in addition to direct carbon sequestration. For example, using natural solutions for erosion control, flood mitigation, and stormwater management might lessen the demand for carbon-intensive infrastructure when incorporating NbS into urban development. By developing more environmentally friendly transport systems, such as green corridors that promote bicycling and walking and lessen dependency on fossil fuels, NbS can also increase the general efficiency of cities. Cities may build synergies between environmental protection, climate mitigation, and urban liveability by incorporating NbS into urban design and policy. This makes cities essential to the shift toward carbon-neutral urban environments (Fankhauser et al., 2022; Möslinger et al., 2023)

iv. Challenges and Opportunities in Pursuing Carbon Neutrality
NbS have a significant potential for sequestering carbon, but their efficacy is limited by issues including land competition and degradation; deforestation alone releases about 4.3 gigatons of CO_2 per year (Rosales et al., 2021). However, there is a significant chance to achieve carbon neutrality because massive afforestation and reforestation programs have the potential to sequester up to 205 gigatons of CO_2 over the course of the next century (Bastin et al., 2019). Urban areas have many obstacles in the pursuit of carbon neutrality, such as the need for large financial commitments, technological advancements, and political will. Large-scale adjustments to the city's energy infrastructure, transit system, and building codes are necessary to achieve carbon neutrality, although doing so can be expensive and difficult. Cities also frequently encounter obstacles like conflicting property uses, regulatory restrictions, and opposition from parties who could be impacted by the changes. A further layer of complexity to reaching carbon neutrality is ensuring that the transition is inclusive and equitable, especially for

marginalized communities who may be disproportionately impacted by climate policy (Zhou, 2023). The goal of carbon neutrality offers many opportunities despite these difficulties. For instance, switching to renewable energy may open new business prospects and spur the development of new technologies as well as employment chances in the green energy industry. Furthermore, by encouraging active transportation and lowering air pollution, carbon neutrality can enhance public health. Cities that make the shift to carbon neutrality effectively can also market themselves as pioneers in sustainability, drawing talent, capital, and visitors. Furthermore, including NbS in urban planning can improve urban resilience, enhancing both the quality of life for citizens and the ability of cities to withstand the effects of climate change. The difficulties in reaching carbon neutrality can ultimately be overcome by creativity, teamwork, and a dedication to long-term sustainability goals (Ulpiani et al., 2023; Tainio et al., 2021).

v. Synergies between NbS, Urban Resilience, and Carbon Neutrality

In the United States alone, urban trees may trap up to 25.6 million metric tons of CO_2 yearly, demonstrating how NbS promote carbon neutrality while enhancing urban resilience (Bherwani et al., 2024). Green infrastructure, such wetlands and vegetated surfaces, helps with climate adaptation and mitigation by reducing flooding and absorbing up to 375 grams of CO_2 per square meter annually (Halim et al., 2022). Resilience and carbon neutrality can work in tandem when NbS is incorporated into urban planning through the installation of green roofs and trees, which can lower city temperatures by up to 4°C. This lowers energy consumption and indirectly lowers emissions (Hayes et al., 2022). Urban resilience, carbon neutrality, and NbS have a lot in common, which makes them complementary approaches to building sustainable cities. By improving cities' capacity to endure and recover from climate-related calamities like heat waves, storms, and floods, NbS contributes to urban resilience. Urban green spaces, for instance, provide the dual benefits of mitigating heat waves and lowering the risk of flooding by absorbing excess rainwater. These resilience advantages are directly related to carbon neutrality since they lessen the need for carbon-intensive interventions, such as air conditioning or large drainage systems, which lowers total carbon emissions (Seddon et al., 2021; Möslinger et al., 2023). Incorporating NbS into urban planning also helps attain carbon neutrality by encouraging sustainable and low-carbon development methods. For example, green infrastructure, such as parks and urban trees, not only improves the liveability and visual appeal of cities but also sequesters carbon, which in turn encourages more sustainable lifestyles. A comprehensive approach to urban sustainability is aided by the numerous co-benefits that NbS offers, including the preservation of biodiversity, better air and water quality, and improved public health. Cities can develop ecosystems that are more fair, sustainable, flexible, and better equipped to handle the problems posed by climate change by integrating NbS with goals for urban resilience and carbon neutrality (Cui et al., 2021; Ranjan, 2024).

4.2.7 Policy and Governance for NbS in Urban Contexts

i. Policy Frameworks Supporting NbS
Policy frameworks are essential for integrating NbS into urban development and planning, offering the strategic, legal, and regulatory direction required to guarantee their execution. Strong national and municipal policies that acknowledge the importance of natural ecosystems in reducing climate change, boosting urban resilience, and promoting sustainable development are usually included in a comprehensive policy framework for NbS. For instance, NbS is specifically endorsed as a crucial instrument in addressing biodiversity loss and climate adaptation in the European Union's Biodiversity Strategy for 2030. Similar to this, national programs stress the incorporation of natural capital and green infrastructure into urban development, such as the UK's 25-Year Environment Plan. Through the establishment of performance measures to track progress, funding methods, and explicit targets, these policies foster the adoption of NbS (Verma et al., 2021; Murti et al., 2020). Locally, NbS can be included in land-use rules and urban planning policies by promoting the integration of green infrastructure into new developments, the protection of green spaces, and the restoration of natural habitats. For example, a lot of towns have "green building codes" that require the use of NbS in construction projects, such as permeable pavements and green roofs. Furthermore, NbS can be supported by urban regulations that offer grants or tax advantages to developers who use sustainable methods. These policy frameworks must be comprehensive, balancing environmental objectives with social and economic concerns, and adaptable enough to consider shifting conditions and local settings to be effective (Pauleit et al., 2020; Biswal et al., 2022).

ii. Governance Structures for Implementing NbS
The implementation of NbS in urban settings requires effective governance structures since they dictate decision-making processes, participant participation, and resource allocation. Many stakeholders, including government agencies, business partners, non-governmental organizations, and community groups, are usually involved in a well-designed NbS governance structure. These stakeholders all have a place to play in the development, execution, and upkeep of NbS projects. With the help of multiple stakeholders, it is ensured that the various advantages of NbS are acknowledged and that projects are customized to fit the requirements of various urban areas. This might entail the creation of specialized NbS units inside local governments, with the responsibility of arranging interdepartmental activities, obtaining funds, and guaranteeing adherence to pertinent laws and guidelines (Murti et al., 2020; Kiss et al., 2022). Moreover, NbS governance frameworks need to make it easier for local, regional, and national governments to work together so that actions and policies are coordinated and reinforce one another. For instance, regional organizations may coordinate efforts across municipal lines to manage common resources like watersheds or coastal areas, while local governments may carry out NbS projects in

accordance with national climate adaptation strategies. To guarantee that NbS initiatives are not only technically sound but also socially just and broadly supported, good governance frameworks should also include procedures for accountability, transparency, and public participation. To enable communities to actively participate in the decision-making process, it is necessary to develop clear channels for communication and feedback (Cui et al., 2021; Hariram et al., 2023).

iii. Multi-Level Governance and NbS
Since multi-level governance entails coordinating actions at the local, regional, national, and even worldwide levels of government, it is essential to the implementation of NbS. This method acknowledges that while NbS are frequently applied locally, in order for them to be successful and long-lasting, higher levels of government must support them. Urban forests and green roofs are two examples of green infrastructure projects that a city may undertake; however, the success of these projects frequently hinges on national policies that supply finance, technical assistance, and regulatory support. In addition, regional governance frameworks can be crucial in overseeing resources that cut across borders, like coastal zones or river basins, and guarantee that NbS are applied consistently and cohesively (Hariram et al., 2023; Beck et al., 2022). International agreements and frameworks have an impact on NbS governance as well, especially when it comes to global climate goals and biodiversity protection objectives. The Convention on Biological Diversity and the Paris Agreement, for instance, set broad goals that direct regional and national NbS policy. Thus, multi-level governance facilitates the harmonization of local efforts with global goals, generating synergies that augment NbS's overall efficacy. In addition to encouraging innovation and expanding the scope of fruitful NbS projects, this strategy makes it easier for various levels of governance to exchange best practices, resources, and expertise. To handle the complexity that results from overlapping authorities and interests, multi-level governance, however, needs clear communication routes, roles and responsibilities that are well defined, and procedures for resolving conflicts (Seddon et al., 2021; Fankhauser et al., 2022).

iv. Stakeholder Engagement and Public Participation
The success of NbS in urban contexts is largely dependent on stakeholder involvement and public participation, which guarantee that the various needs and viewpoints of the community are considered during the planning and implementation phases. Involving stakeholders, such as nearby citizens, companies, non-governmental organizations, and educational institutions, promotes community ownership and accountability while aiding in the development of consensus around NbS projects. To guarantee that the advantages of NbS are shared fairly, effective stakeholder involvement entails open communication, inclusive decision-making procedures, and the participation of marginalized groups. To customize NbS projects to local contexts and priorities, for example, input from diverse stakeholders can be gathered through workshops, public consultations, and collaborative

planning sessions (Murti et al., 2020; Kiss et al., 2022). The public's involvement extends beyond simple advice; It entails actively participating in NbS project design, execution, and upkeep. By ensuring that projects are not only technically sound but also socially and culturally relevant, this participatory method can improve the sustainability of NbS. Public participation in NbS programs can also strengthen the bond between urban dwellers and the wild world and increase understanding of the value of ecosystem services. In actuality, cities can set up community stewardship initiatives to care for and keep an eye on green infrastructure, or they can adopt participatory budgeting procedures that give citizens a direct vote in how money for NbS is distributed. Cities can ensure the long-term viability of NbS by enhancing its resilience and adaptation through the active participation of local populations (Murti et al., 2020; Kiss et al., 2022).

v. Financing Mechanisms for NbS in Cities

Financing mechanisms are a critical aspect of implementing NbS in urban areas since they offer the resources required to plan, create, and sustain innovative solutions over time. Conventional funding sources, such as public grants and government budgets, are still crucial, but they frequently fall short of covering the high expenses involved in large-scale NbS initiatives. Consequently, cities are progressively delving into novel funding methods such as payments for ecosystem services (PES), green bonds, and PPPs. Green bonds, for instance, draw investors interested in sustainable development and enable towns to acquire money expressly for environmental initiatives such as NbS. PPPs can help implement NbS by utilizing the resources and experience of the private sector, all the while ensuring that initiatives meet public objectives and yield long-term benefits (Black et al., 2021; Ulpiani et al., 2023). Furthermore, NbS financing may entail methods that quantify the ecological services' economic worth. For instance, PES programs can pay landowners or communities to preserve ecosystems that provide advantages to the public, such as carbon sequestration or clean water. This strategy encourages the preservation and repair of natural habitats while simultaneously giving NbS a source of income. Cities might also look at value capture finance, which uses the boost in property prices brought about by NbS projects—like restoring shoreline areas or creating green spaces—to pay for the initial expenditure. By creating a varied range of funding options, communities can lessen their need for a single funding source, distributing the risk and guaranteeing the long-term financial viability of NbS projects (Brownson et al., 2020; Ulpiani et al., 2023).

Beyond these mechanisms, additional financial strategies are emerging to support NbS implementation while aligning with broader carbon neutrality objectives.

a. Carbon Credits and NbS for Urban Carbon Neutrality

The incorporation of urban NbS projects into carbon markets via carbon credits is one of the most promising methods for funding NbS. NbS initiatives, like green roofs, wetland restoration, and urban afforestation, can produce quantifiable carbon sequestration benefits under carbon offset schemes. These benefits are then offered as carbon credits to companies, governments, or organizations looking to lessen their emissions. Programs like the Gold Standard and the Verified Carbon Standard offer frameworks for attesting to the carbon sequestration from urban greening projects, enabling cities to draw in investment from businesses that have made net-zero commitments (Croci & Lucchitta, 2022; Pan et al., 2023; Johnstone et al., 2023).

b. Tax Incentives and Green Infrastructure Mandates
For projects that use green infrastructure, such as permeable pavements, bioswales, and green rooftops, governments might incentivize NbS investments by offering tax credits, subsidies, or lower property taxes. To encourage the use of NbS, several cities offer stormwater fee reductions to property owners who install nature-based stormwater management systems. In a similar vein, NbS improvements can raise property values, which tax increment financing can then deploy in other sustainability projects. (Thompson et al., 2023; Knapik et al., 2024).

c. Green Bonds and Sustainability-Linked Loans
Climate resilience funds, sustainability-linked loans, and green bonds can all offer specific funding for NbS initiatives. Municipalities and private organizations can collect money for initiatives like urban forestry, coastal resilience, and climate-adaptive urban design thanks to these financial instruments, which draw investors that place a high value on sustainability. Sustainability-linked loans are a useful instrument for promoting NbS integration since they modify interest rates in response to the borrower's accomplishment of predetermined sustainability goals (Brears, 2022).

d. Payments for Ecosystem Services and Biodiversity Credits
PES systems, in addition to carbon credits, allow communities, corporations, or governments to reimburse landowners for ecosystem services rendered by NbS initiatives. Flood prevention, urban cooling, biodiversity preservation, and air quality enhancement are a few examples. New biodiversity credit markets work similarly to carbon markets, rewarding ecological interconnectedness and habitat restoration initiatives, which further encourage the use of NbS in urban design (Biswal et al., 2022; Thompson et al., 2023).

e. Regulatory Mechanisms and NbS-Driven Urban Policies
Mandatory green space requirements, net-zero construction codes with NbS components, and climate adaptation laws that give natural infrastructure precedence over conventional gray infrastructure are some examples of policies that include NbS into urban carbon neutrality frameworks. In order to guarantee that NbS is incorporated into urban expansion plans, cities like Singapore and Copenhagen have put in place policy frameworks that require green coverage in new construction (Mishra et al., 2024; Tun et al., 2024).

In conclusion, cities must use a mix of financial tools and policy-driven incentives to successfully grow NbS and promote urban carbon neutrality. A hybrid strategy that makes use of carbon markets, green bonds, tax breaks, PES plans, and regulatory requirements can build a robust financial framework that guarantees the long-term viability of NbS projects. Cities may advance sustainability and resilience goals and open up new revenue streams by coordinating NbS finance with climate action goals.

4.2.8 Future Directions and Innovations in NbS for Smart Cities

i. Emerging Trends and Technologies
The implementation and management of NbS by cities to tackle urban difficulties associated with climate change, sustainability, and resilience are undergoing a rapid transformation propelled by emerging trends and technology. A prominent development in NbS planning and management is the incorporation of digital technologies, including big data analytics, AI, and the IoT. More accurate and flexible management of green infrastructure is made possible by these technologies, which allow for real-time monitoring and data collection on a variety of environmental parameters, including biodiversity levels, soil moisture, and air quality. Urban green areas can benefit from the continuous data on plant health and water requirements that smart sensors installed in them can offer, which can optimize irrigation and minimize water consumption. Further, Planners can create more effective interventions by modeling the effects of NbS on urban microclimates with the use of AI and machine learning technologies (Pedral Sampaio et al., 2022; Matasov et al., 2020). Using mobile applications and digital platforms to involve residents in NbS monitoring and maintenance is another new trend. These methods facilitate community participation in environmental stewardship by enabling residents to report issues, including pollution or invasive species, directly to city officials. Additionally, NbS advantages can be promoted and sustainable practices like water and waste conservation can be encouraged using digital platforms. The increasing interest in digital twins reflects this trend toward digitalization as well—virtual replicas of physical environments—that enables cities to model how NbS will react under various situations, such as severe weather or population increase. NbS is becoming more effective and efficient thanks to these digital advances, which are also improving their scalability and resilience in the face of rapidly increasing urbanization and climate change (Pedral Sampaio et al., 2022; Anderson et al., 2023).

ii. Innovations in NbS for Urban Resilience
Innovations in NbS are playing a critical role in enhancing urban resilience to climate change by addressing the vulnerabilities of cities to extreme weather events, such as floods, heatwaves, and storms. The creation of multifunctional green infrastructure that blends conventional NbS with contemporary engineering methods is one area of innovation. For example, hybrid solutions like "sponge cities" combine sophisticated drainage systems with natural elements like wetlands and permeable pavements to better

absorb and regulate stormwater and lower the danger of urban flooding. By strengthening urban ecosystems' ability to adapt, these solutions not only lessen the short-term effects of climate-related calamities but also promote long-term resilience (Costello et al., 2024). The use of adaptable and modular designs that are simple to scale or change in response to shifting urban demands or environmental conditions is another important breakthrough in NbS. For instance, existing buildings can have vertical gardens or modular green roofs added to them to help with cooling during heatwaves, enhance air quality, and support urban biodiversity. These modular systems are highly adaptive to upcoming issues since they can be expanded or altered in response to changes in the needs of the city. Furthermore, cutting-edge methods of urban forestry, like the Miyawaki method, which produces dense, quickly expanding native forests, are being used to improve urban green spaces quickly and offer instant advantages in the form of temperature regulation, habitat creation, and carbon sequestration. These developments show that NbS may be scalable and useful, providing cities with an adaptable toolset for enhancing resilience in the face of climate unpredictability (Costello et al., 2024).

iii. The Future of Smart Cities in the Context of Climate Change

The need to address the difficulties posed by climate change is increasingly shaping the future of smart cities, and NbS are becoming an essential component of this evolution. The integration of NbS is currently the main emphasis of smart cities, which use data and technology to improve urban living. This will increase liveability, resilience, and sustainability. Accordingly, it is anticipated that smart cities will optimize NbS design, implementation, and management using AI, IoT, and advanced data analytics. For instance, with the analysis of data from multiple sources, including weather forecasts, environmental sensors, and satellite images, smart cities may create predictive models that direct the deployment of NbS strategically to optimize its effects on flood management, heat mitigation, and air quality improvement. Through proactive adaptation to future climatic conditions and patterns of urban growth, this strategy guarantees that NbS are not just reactive but also proactive (Kumar et al., 2021; Petit, 2022; Bibri, 2022). The notion of a "smart city" is expected to shift as cities increasingly deal with the combined challenges of growing urbanization and climate change. In this paradigm, technology and the natural world collaborate to build resilient, carbon-neutral urban settings. It is anticipated that "green digital twins"—virtual models including NbS and simulating their performance under various development and climate scenarios—will be given top priority in future smart cities. By assisting politicians and urban planners in making well-informed decisions on infrastructure expenditures, land use, and disaster preparedness, these digital twins can guarantee that NbS is a key component of the city's climate adaptation and mitigation plans. Furthermore, citizen-centric strategies will probably be used in future smart cities. These strategies will involve including citizens in the co-creation and stewardship of NbS using digital platforms, which will strengthen the bond between urban areas and their surrounding environment (García Fernández & Peek, 2020; Kiss et al., 2022).

iv. Potential for Scaling Up NbS

There is a great chance for NbS to be scaled up as more people become aware of its potential to address urban problems and offer other co-benefits, such as social well-being, biodiversity protection, and climate mitigation. A focused effort needs to be made in several areas, including policy, public participation, technology, financing, and NbS scalability. In terms of policy, governments can facilitate scaling by ensuring that NbS is incorporated into national and local urban planning frameworks and that these solutions are given precedence when decisions are being made on land use, infrastructure development, and climate adaptation plans. For instance, mandating green infrastructure for newly constructed buildings or providing financial incentives for NbS retrofits in already-existing cities can significantly boost their adoption (Cortinovis et al., 2022; Scolobig et al., 2023). Due to the high cost of large-scale implementation, financial channels are also essential for scaling NbS. The resources required for NbS projects can be mobilized through creative financing strategies like impact investment, PPPs, and green bonds. Furthermore, by including NbS in risk management and insurance frameworks, firms and localities can get financial incentives to use these solutions and lessen their exposure to climate-related hazards. Advancements in technology, like digital platforms, remote sensing, and GIS mapping, can facilitate more effective planning, monitoring, and management, which in turn can help NbS scale. Furthermore, local communities and stakeholders' involvement is crucial to the effective scaling of NbS since it guarantees that solutions are contextually relevant, culturally appropriate, and supported by those who stand to benefit the most. By addressing these key factors, cities can significantly expand the reach and impact of NbS, making them a cornerstone of urban sustainability and resilience efforts (Kumar et al., 2021; Kiss et al., 2022; Ulpiani et al., 2023).

4.3 CONCLUSION

This chapter has explored the critical role that NbS can play in enhancing urban resilience to climate-related disasters while contributing to the overarching goals of sustainability and carbon neutrality. The analysis shows that NbS, like wetlands restoration, urban green spaces, and green infrastructure, have the potential to improve social well-being, improve air quality, and increase biodiversity in addition to being effective at reducing the effects of extreme weather events. Data-driven methods can optimize NbS management and efficacy in the integration of smart city technology, which has been recognized as a potent synergy. According to the findings, it is crucial to see NbS as a multipurpose strategy that combines ecological sustainability with goals for urban development, thereby increasing cities' capacity to respond to the effects of climate change.

Several legislative suggestions have surfaced to fully utilize NbS in the construction of carbon-neutral and climate-resilient towns. Firstly, NbS must be incorporated into urban planning frameworks, making sure that they are viewed as fundamental elements of urban infrastructure rather than as optional or auxiliary elements. To achieve this, regulations requiring the incorporation of green areas, sustainable water management practices, and other NbS into both new and existing urban developments must be developed. Financial incentives and regulatory support must be extended to incentivize PPPs to allocate resources toward NbS. Thirdly, to improve the efficacy and efficiency of NbS, regulations ought to encourage the use of smart technology for management and monitoring. Lastly, community involvement and public awareness are critical; regulations need to incorporate measures for citizen education and involvement in the co-creation and upkeep of NbS, promoting a sense of ownership and responsibility toward these green initiatives.

A viable way ahead is the integration of NbS into the smart city paradigm, as cities increasingly confront the challenges posed by climate change. To achieve sustainability, resilience, and carbon neutrality, NbS will need to be embraced as a cornerstone of urban development going forward, surpassing traditional gray infrastructure. Developing cross-sectoral collaboration, scaling up successful NbS models, and utilizing cutting-edge technology to optimize their deployment and management are the steps toward the future. Moreover, thorough frameworks that assess the efficacy of NbS in terms of both environmental impact and socioeconomic advantages must be developed. The incorporation of NbS into urban ecosystems will be essential to maintaining cities' ability to function as liveable, flexible, and environmentally friendly places as they expand. Cities may steer toward a sustainable future where urban resilience and carbon neutrality are not just goals but achievable realities by giving priority to NbS.

REFERENCES

Abbass, K., Qasim, M. Z., Song, H., Murshed, M., Mahmood, H., & Younis, I. (2022). A review of the global climate change impacts, adaptation, and sustainable mitigation measures. *Environmental Science and Pollution Research*, 29(28), 42539–42559.

Abedi, M., Fangueiro, R., Correia, A. G., & Shayanfar, J. (2023). Smart geosynthetics and prospects for civil infrastructure monitoring: A comprehensive and critical review. *Sustainability*, 15(12), 9258.

Adams, C., Frantzeskaki, N., & Moglia, M. (2023). Space for mainstreaming? Learning from the implementation of urban forest strategies in metropolitan Melbourne. *Australian Planner*, 59(2), 154–169.

Agboola, O. P., & Tunay, M. (2023). Urban resilience in the digital age: The influence of Information-Communication technology for sustainability. *Journal of Cleaner Production*, 428, 139304.

Ahmed, I., Asif, M., Alhelou, H. H., & Khalid, M. (2024). A review on enhancing energy efficiency and adaptability through system integration for smart buildings. *Journal of Building Engineering*, *89*, 109354.

Al-Kodmany, K. (2024). Promoting health in dense cities through vertical greeneries: The case of plant-and tree-covered tall buildings. In P. Du, K. Al-Kodmany, & M. M. Ali (Eds.), *The Routledge handbook on greening high-density cities* (pp. 509–529). Routledge.

Alagirisamy, B., & Ramesh, P. (2022). Smart sustainable cities: Principles and future trends. In I. Pal & S. Kolathayar (Eds.), *Sustainable cities and resilience: Select proceedings of VCDRR 2021* (pp. 301–316). Springer Singapore.

Alahi, M. E. E., Sukkuea, A., Tina, F. W., Nag, A., Kurdthongmee, W., Suwannarat, K., & Mukhopadhyay, S. C. (2023). Integration of IoT-enabled technologies and artificial intelligence (AI) for smart city scenario: Recent advancements and future trends. *Sensors*, 23(11), 5206.

Ali, M., Javeed, H. M. R., Tariq, M., Qamar, R., Deeba, F., Ihsan, M. Z., … & Mubeen, M. (2023). Climate and development. In W. N. Jatoi, M. Mubeen, M. Zaffar Hashmi, S. Ali, S. Fahad, & K. Mahmood (Eds.), *Climate change impacts on agriculture: Concepts, issues and policies for developing countries* (pp. 191–208). Springer International Publishing.

Ali, G., Mijwil, M. M., Adamopoulos, I., & Ayad, J. (2025). Leveraging the internet of things, remote sensing, and artificial intelligence for sustainable forest management. *Babylonian Journal of Internet of Things*, *2025*, 1–65.

Aljashaami, B. A., Ali, B. M., Salih, S. A., Alwan, N. T., Majeed, M. H., Ali, O. M., … & Shcheklein, S. E. (2024). Recent improvements to heating, ventilation, and cooling technologies for buildings based on renewable energy to achieve zero-energy buildings: A systematic review. *Results in Engineering*, *23*, 102769.

Almeida, A. P., Liberalesso, T., Silva, C. M., & Sousa, V. (2023). Combining green roofs and rainwater harvesting systems in university buildings under different climate conditions. *Science of the Total Environment*, 887, 163719.

Alsamhi, S. H., Afghah, F., Sahal, R., Hawbani, A., Al-qaness, M. A., Lee, B., & Guizani, M. (2021). Green internet of things using UAVs in B5G networks: A review of applications and strategies. *Ad Hoc Networks*, 117, 102505.

Amirzadeh, M., Sobhaninia, S., & Sharifi, A. (2022). Urban resilience: A vague or an evolutionary concept? *Sustainable Cities and Society*, 81, 103853.

Anderson, V., Suneja, M., & Dunjic, J. (2023). Sensing and measurement techniques for evaluation of nature-based solutions: A state-of-the-art review. *Land*, 12(8), 1477.

Andrae, S. (2024). The establishment of a gross ecosystem product: A country comparison. In S. Boubaker & T.-H. Le (Eds.), *Green finance and sustainable development goals* (pp. 369–407). World Scientific Publishing Co. Pte. Ltd.

Anguelovski, I., Zografos, C., Klause, K. A., & Connolly, J. J. (2022). Urban transformational adaptation: Contestation and struggles for authority in the pilot barcelona superblock of poblenou. In I. Ruiz-Mallén, H. March, & M. Satorras (Eds.), *Urban resilience to the climate emergency: Unravelling the transformative potential of institutional and grassroots initiatives* (pp. 65–91). Springer International Publishing.

Anguelovski, I., Honey-Rosés, J., & Marquet, O. (2023). Equity concerns in transformative planning: Barcelona's Superblocks under scrutiny. *Cities & Health*, 7(6), 950–958.

Antal, H., & Bhutani, S. (2023). Identifying linkages between climate change, urbanization, and population ageing for understanding vulnerability and risk to older people: A review. *Ageing International*, 48(3), 816–839.

Ariyarathna, I. S., Abeyrathna, W. P., Jamei, E., & Chau, H. W. (2023). A review of the application of blue–green infrastructure (BGI) as an effective urban flood mitigation strategy for livable and healthy cities in Australia. *Architecture*, 3(3), 461–476.

Baker, N. T., Sullivan, D. J., Selbig, W. R., Haefner, R., Lampe, D. C., Bayless, E. R., & McHale, M. R. (2022). Green infrastructure in the Great Lakes—Assessment of performance, barriers, and unintended consequences (No. 1496). US Geological Survey.

Balk, D., McPhearson, T., Cook, E. M., Knowlton, K., Maher, N., Marcotullio, P., … & Wagner, G. (2024). NPCC4: Concepts and tools for envisioning New York City's futures. *Annals of the New York Academy of Sciences*, *1539*(1), 277–322.

Barker, A., Garcia-Blanco, G., Garcia, I., & Aguirre-Such, A. (2024). The role of strategic planning in Nature-based Solutions (NBS) transformation: An evaluation of the Green Cities Framework in mainstreaming NBS in 6 European countries. *Nature-Based Solutions*, 6, 100157.

Baskar, P., Khang, A., Annadurai, S., Amritha, S. S., & Annadurai, P. (2024). Quantum computing smart city: Fostering sustainability and green infrastructure with smartcCity. In A. Khang (Ed.), *Applications and principles of quantum computing* (pp. 162–213). IGI Global.

Bastin, J.-F., Finegold, Y., Garcia, C., Mollicone, D., Rezende, M., Routh, D., Zohner, C. M., & Crowther, T. W. (2019). The global tree restoration potential. *Science*, 365(6448), 76–79. https://doi.org/10.1126/science.aax0848

Battemarco, B. P., Tardin-Coelho, R., Veról, A. P., de Sousa, M. M., da Fontoura, C. V. T., Figueiredo-Cunha, J., … & Miguez, M. G. (2022). Water dynamics and blue-green infrastructure (BGI): Towards risk management and strategic spatial planning guidelines. *Journal of Cleaner Production*, 333, 129993.

Beck, M. W., Heck, N., Narayan, S., Menéndez, P., Reguero, B. G., Bitterwolf, S., … & Losada, I. J. (2022). Return on investment for mangrove and reef flood protection. *Ecosystem Services*, 56, 101440.

Beitelmal, W. H., Nwokolo, S. C., Meyer, E. L., & Ahia, C. C. (2024). Exploring adaptation strategies to mitigate climate threats to transportation infrastructure in Nigeria: Lagos City, as a case study. *Climate*, 12(8), 117.

Bevilacqua, P. (2021). The effectiveness of green roofs in reducing building energy consumptions across different climates. A summary of literature results. *Renewable and Sustainable Energy Reviews*, 151, 111523.

Bherwani, H., Banerji, T., & Menon, R. (2024). Role and value of urban forests in carbon sequestration: Review and assessment in Indian context. *Environment, Development and Sustainability*, 26(1), 603–626.

Bhushan, B., & Sharma, A. (2022). Sea-level rise due to climate change. In S. Eslamian & F. Eslamian (Eds.), *Flood handbook* (pp. 265–284). CRC Press.

Bibri, S. E. (2022). Eco-districts and data-driven smart eco-cities: Emerging approaches to strategic planning by design and spatial scaling and evaluation by technology. *Land Use Policy*, 113, 105830.

Bibri, S. E., Krogstie, J., & Kärrholm, M. (2020). Compact city planning and development: Emerging practices and strategies for achieving the goals of sustainability. *Developments in the Built Environment*, 4, 100021.

Biswal, B. K., Bolan, N., Zhu, Y. G., & Balasubramanian, R. (2022). Nature-based Systems (NbS) for mitigation of stormwater and air pollution in urban areas: A review. *Resources, Conservation and Recycling*, 186, 106578.

Black, M. S., Parry, I., Roaf, M. J., & Zhunussova, K. (2021). *Not yet on track to net zero: The urgent need for greater ambition and policy action to achieve Paris temperature goals.* International Monetary Fund.

Boateng, E. A., Asibey, M. O., Cobbinah, P. B., Adutwum, I. O., & Blija, D. K. (2023). Enabling nature-based solutions: Innovating urban climate resilience. *Journal of Environmental Management*, 332, 117433.

Bona, S., Silva-Afonso, A., Gomes, R., Matos, R., & Rodrigues, F. (2022). Nature-based solutions in urban areas: A European analysis. *Applied Sciences*, 13(1), 168.

Boulton, A., Dahm, C., Correa, L., Kingsford, R., Jenkins, K., Negishi, J., … & Goodwin, P. (2013). Good news: Progress in successful river conservation and restoration. In A. Elosegi & S. Sabater (Eds.), *River conservation: Challenges and opportunities.* Fundación BBVA.

Brears, R. C. (2022). Green bonds, loans, credit lines, and microfinance financing nature-based solutions. In R. C. Brears (Ed.), *Financing nature-based solutions: Exploring public, private, and blended finance models and case studies* (pp. 105–134). Springer International Publishing. https://doi.org/10.1007/978-3-030-93325-8

Bressane, A., Loureiro, A. I. S., de Castro Medeiros, L. C., Moruzzi, R., Formiga, J. K. S., Negri, R. G., & Saraiva, A. C. V. (2024). Intelligent nature-based solutions in the 1st smart sustainable Brazilian City: Insights and lessons learned. *Nature-Based Solutions*, 6, 100161.

Brownson, K., Anderson, E. P., Ferreira, S., Wenger, S., Fowler, L., & German, L. (2020). Governance of payments for ecosystem services influences social and environmental outcomes in Costa Rica. *Ecological Economics*, 174, 106659.

Bueb, B., Tröltzsch, J., Reichwein, D., Oldenburg, C., Favero, F., Bodle, R., ... & Voß, M. (2021). Towards sustainable adaptation pathways. A concept for integrative actions to achieve the 2030 Agenda, Paris Agreement and Sendai Framework. Conceptual and analytical paper (No. UBA-FB--000590/3, ENG). Umweltbundesamt (UBA).

Bulkeley, H., Almassy, D., Fransen, A., Maia, S., & Toxopeus, H. (2023). *Enhancing urban nature provision in the Netherlands*. PBL Netherlands Environmental Assessment Agency, PBL publication, (5022).

Carvalho, P. N., Finger, D. C., Masi, F., Cipolletta, G., Oral, H. V., Tóth, A., ... & Exposito, A. (2022). Nature-based solutions addressing the water-energy-food nexus: Review of theoretical concepts and urban case studies. *Journal of Cleaner Production*, 338, 130652.

Castelo, S., Amado, M., & Ferreira, F. (2023). Challenges and opportunities in the use of nature-based solutions for urban adaptation. *Sustainability*, 15(9), 7243.

Chapagain, K., Babel, M. S., Karthe, D., & Stamm, J. (2024). Integrated assessment of water–energy–food nexus: Conceptual framework and application to the Ping River basin, Thailand. *International Journal of Water Resources Development*, 40(2), 284–318.

Choudhary, B., Dhar, V., & Pawase, A. S. (2024). Blue carbon and the role of mangroves in carbon sequestration: Its mechanisms, estimation, human impacts and conservation strategies for economic incentives. *Journal of Sea Research*, 199, 102504.

Chowdhuri, A., & Gupta, C. K. (2023). Gauging the assessment of some anthropogenic factors driving climate-change. *Current World Environment*, 18(2), 614.

Clarke, B., Otto, F., Stuart-Smith, R., & Harrington, L. (2022). Extreme weather impacts of climate change: An attribution perspective. *Environmental Research: Climate*, 1(1), 012001.

Clarke, M., Cadaval, S., Wallace, C., Anderson, E., Egerer, M., Dinkins, L., & Platero, R. (2023). Factors that enhance or hinder social cohesion in urban greenspaces: A literature review. *Urban Forestry & Urban Greening*, 84, 127936.

Cong, C., Pan, H., Page, J., Barthel, S., & Kalantari, Z. (2023). Modeling place-based nature-based solutions to promote urban carbon neutrality. *Ambio*, 52(8), 1297–1313.

Cortinovis, C., Olsson, P., Boke-Olén, N., & Hedlund, K. (2022). Scaling up nature-based solutions for climate-change adaptation: Potential and benefits in three European cities. *Urban Forestry & Urban Greening*, 67, 127450.

Costello, M. M., West, D., Szydlowski, S., & Ramirez, B. (2024). Global health impact from climate change. *Clinical Social Work and Health Intervention*, *15*(5), 16–20.

Croci, E., Lucchitta, B. (2022). Climate Change and Urban Nature: Impacts and Policies at the Urban Level. In: Kim, KG., Thioye, M. (eds) *Planning Climate Smart and Wise Cities. The Urban Book Series. Springer*, Cham. https://doi.org/10.1007/978-3-030-80165-6_3

Croeser, T., Bekessy, S. A., Garrard, G. E., & Kirk, H. (2024). Nature-based solutions for urban biodiversity: Spatial targeting of retrofits can multiply ecological connectivity benefits. *Landscape and Urban Planning*, 251, 105169.

Cui, M., Ferreira, F., Fung, T. K., & Matos, J. S. (2021). Tale of two cities: How nature-based solutions help create adaptive and resilient urban water management practices in Singapore and Lisbon. *Sustainability*, 13(18), 10427.

D'Evelyn, S. M., Jung, J., Alvarado, E., Baumgartner, J., Caligiuri, P., Hagmann, R. K., ... & Spector, J. T. (2022). Wildfire, smoke exposure, human health, and environmental justice need to be integrated into forest restoration and management. *Current Environmental Health Reports*, 9(3), 366–385.

Datola, G. (2023). Implementing urban resilience in urban planning: A comprehensive framework for urban resilience evaluation. *Sustainable Cities and Society*, 98, 104821.

Debele, S. E., Leo, L. S., Kumar, P., Sahani, J., Ommer, J., Bucchignani, E., … & Di Sabatino, S. (2023). Nature-based solutions can help reduce the impact of natural hazards: A global analysis of NBS case studies. *Science of the Total Environment*, 902, 165824.

Dianat, H., Wilkinson, S., Williams, P., & Khatibi, H. (2022). Choosing a holistic urban resilience assessment tool. *International Journal of Disaster Risk Reduction*, 71, 102789.

Dickinson, D. C., & Ramalho, C. E. (2022). A balancing act: Biodiversity and human wellbeing considerations in the management of urban forest in a global biodiversity hotspot. *Urban Forestry & Urban Greening*, 74, 127656.

Dickey, A., Kosovac, A., Fastenrath, S., Acuto, M., & Gleeson, B. (2022). Fragmentation and urban knowledge: An analysis of urban knowledge exchange institutions. *Cities*, 131, 103917.

Domke, G. M., Walters, B. F., Nowak, D. J., Greenfield, E. J., Smith, J. E., Nichols, M. C., … & Wirth, T. C. (2022). Greenhouse gas emissions and removals from forest land, woodlands, urban trees, and harvested wood products in the United States, 1990–2020. Resource Update FS–382.US Department of Agriculture, Forest Service, Northern Research Station., 382.

Dorst, H., van der Jagt, A., Toxopeus, H., Tozer, L., Raven, R., & Runhaar, H. (2022). What's behind the barriers? Uncovering structural conditions working against urban nature-based solutions. *Landscape and Urban Planning*, 220, 104335.

Eltinay, N., & Egbu, C. (2024). *Urban resilience and climate change in the MENA region*. Taylor & Francis.

Epelde, L., Mendizabal, M., Gutiérrez, L., Artetxe, A., Garbisu, C., & Feliu, E. (2022). Quantification of the environmental effectiveness of nature-based solutions for increasing the resilience of cities under climate change. *Urban Forestry & Urban Greening*, 67, 127433.

Esraz-Ul-Zannat, M., Dedekorkut-Howes, A., & Morgan, E. A. (2024). A review of nature-based infrastructures and their effectiveness for urban flood risk mitigation. *Wiley Interdisciplinary Reviews: Climate Change*, *15*(5), e889.

Fankhauser, S., Smith, S. M., Allen, M., Axelsson, K., Hale, T., Hepburn, C., … & Wetzer, T. (2022). The meaning of net zero and how to get it right. *Nature Climate Change*, 12(1), 15–21.

Farina, G., Le Coent, P., & Hérivaux, C. (2024). Do urban environmental inequalities influence demand for nature-based solutions? *Ecological Economics*, 224, 108298.

Ferreira, C. S., Kalantari, Z., Hartmann, T., & Pereira, P. (Eds.). (2022). *Nature-based solutions for flood mitigation: Environmental and socio-economic aspects (Vol.* 107*)*. Springer Nature.

Ferreira, C. S. S., Potočki, K., Kapović-Solomun, M., & Kalantari, Z. (2021). Nature-based solutions for flood mitigation and resilience in urban areas. In C. S. S. Ferreira, Z. Kalantari, T. Hartmann, & P. Pereira (Eds.), *Nature-based solutions for flood mitigation: Environmental and socio-economic aspects* (pp. 59–78). Springer International Publishing.

Foster, S. R., Baptista, A., Nguyen, K. H., Tchen, J., Tedesco, M., & Leichenko, R. (2024). NPCC4: Advancing climate justice in climate adaptation strategies for New York City. *Annals of the New York Academy of Sciences*. https://doi.org/10.1111/nyas.15148

Ganapathi, H., Awasthi, S., & Vasudevan, P. (2024). Wetlands as a nature-based solution for urban water management. In A. Stefanakis, H. V. Oral, C. Calheiros, & P. Carvalho (Eds.), *Nature-based solutions for circular management of urban water* (pp. 259–275). Springer International Publishing.

García Fernández, C., & Peek, D. (2020). Smart and sustainable? Positioning adaptation to climate change in the European smart city. *Smart Cities*, 3(2), 511–526.

Gies, E. (2022). *Water always wins: Thriving in an age of drought and deluge*. University of Chicago Press.

Gopinadh Garre, S. S. (2024). Climate change and nature-based solutions: Harnessing nature's potential for adaptation and mitigation. In K. K. Sharma, S. Sharma, V. K. Pandey, & R. Singh (Eds.), *Climate change and human adaptation in India: Sustainability and public policy* (pp. 115–135). Springer Nature Switzerland.

Granata, F., & Di Nunno, F. (2025). Pathways for hydrological resilience: Strategies for adaptation in a changing climate. *Earth Systems and Environment*, 1–29. https://doi.org/10.1007/s41748-024-00567-x

Gupta, A., & De, B. (2024). A systematic review on urban blue-green infrastructure in the south Asian region: Recent advancements, applications, and challenges. *Water Science & Technology*, 89(2), 382–403.

Gupta, S. K., Kanga, S., Meraj, G., Singh, S. K., Singh, S., Sajan, B., … & Mishra, V. N. (2024). Optimizing land use for climate mitigation using nature-based solution (NBS) strategy: A study on afforestation potential and carbon sequestration in Rajasthan, India. *Discover Geoscience*, 2(1), 36.

Halim, M. A., Vantellingen, J., Gorgolewski, A. S., Rose, W. K., Drake, J. A., Margolis, L., & Thomas, S. C. (2022). Greenhouse gases and green roofs: carbon dioxide and methane fluxes in relation to substrate characteristics. *Urban Ecosystems*, *25*(2), 1–12.

Hamel, P., & Tan, L. (2022). Blue–green infrastructure for flood and water quality management in Southeast Asia: Evidence and knowledge gaps. *Environmental Management*, 69(4), 699–718.

Han, J., Wang, C., Deng, S., & Lichtfouse, E. (2023). China's sponge cities alleviate urban flooding and water shortage: A review. *Environmental Chemistry Letters*, 21(3), 1297–1314.

Han, S., & Kuhlicke, C. (2021). Barriers and drivers for mainstreaming nature-based solutions for flood risks: The case of South Korea. *International Journal of Disaster Risk Science*, 12, 661–672.

Hanna, E., Bruno, D., & Comín, F. A. (2024). The ecosystem services supplied by urban green infrastructure depend on their naturalness, functionality and imperviousness. *Urban Ecosystems*, 27(1), 187–202.

Hansson, C. B., Lenhart, J., Fahnestock, J., Calder, J., Leire, C., & Ji, W. (2023). *Urban climate governance: Towards 1.5° Celsius alignment*. Lund University.

Hariram, N. P., Mekha, K. B., Suganthan, V., & Sudhakar, K. (2023). Sustainalism: An integrated socio-economic-environmental model to address sustainable development and sustainability. *Sustainability*, 15(13), 10682.

Hayes, A. T., Jandaghian, Z., Lacasse, M. A., Gaur, A., Lu, H., Laouadi, A., ... & Wang, L. (2022). Nature-based solutions (NBSs) to mitigate urban heat island (UHI) effects in Canadian cities. *Buildings*, *12*(7), 925.

He, B. J. (2022). Green building: A comprehensive solution to urban heat. *Energy and Buildings*, 271, 112306.

He, H., & Roulet, N. T. (2023). Improved estimates of carbon dioxide emissions from drained peatlands support a reduction in emission factor. *Communications Earth & Environment*, 4(1), 436.

He, Q., Tapia, F., & Reith, A. (2023). Quantifying the influence of nature-based solutions on building cooling and heating energy demand: A climate specific review. *Renewable and Sustainable Energy Reviews*, 186, 113660.

Hernández-Delgado, E. A. (2024). Coastal restoration challenges and strategies for small Island developing states in the face of sea level rise and climate change. *Coasts*, 4(2), 235–286.

Hull, E. V. (2024). Protecting coastal communities and marine biodiversity under the coastal zone management act. *South Carolina Law Review*, 76, 1.

Huynh, L. T. M., Su, J., Wang, Q., Stringer, L. C., Switzer, A. D., & Gasparatos, A. (2024). Meta-analysis indicates better climate adaptation and mitigation performance of hybrid engineering-natural coastal defence measures. *Nature Communications*, 15(1), 2870.

Jabbar, M., Yusoff, M. M., & Shafie, A. (2022). Assessing the role of urban green spaces for human well-being: A systematic review. *GeoJournal, 87*(5), 1–19.

Jafari, M., Kavousi-Fard, A., Chen, T., & Karimi, M. (2023). A review on digital twin technology in smart grid, transportation system and smart city: Challenges and future. *IEEE Access*, 11, 17471–17484.

Jain, M. (2022). Increasing atmospheric extreme events and role of disaster risk management: Dimensions and approaches. In P. Saxena, A. Shukla, & A. K. Gupta (Eds.), *Extremes in atmospheric processes and phenomenon: Assessment, impacts and mitigation* (pp. 303–328). Springer Nature Singapore.

Jiang, C., Li, J., Hu, Y., Yao, Y., & Li, H. (2022). Construction of water-soil-plant system for rainfall vertical connection in the concept of sponge city: A review. *Journal of Hydrology*, 605, 127327.

Johnstone, I. (2023). Nature-based solutions and the voluntary carbon market: Opportunities and limits. In W. L. Filho, G. J. Nagy, & D. Y. Ayal (Eds.), *Handbook of nature-based solutions to mitigation and adaptation to climate change* (pp. 1–23). Springer International Publishing.

Jones, L., Anderson, S., Læssøe, J., Banzhaf, E., Jensen, A., Bird, D. N., ... & Zandersen, M. (2022). A typology for urban green infrastructure to guide multifunctional planning of nature-based solutions. *Nature-Based Solutions*, 2, 100041.

Jordan, P., & Fröhle, P. (2022). Bridging the gap between coastal engineering and nature conservation? A review of coastal ecosystems as nature-based solutions for coastal protection. *Journal of Coastal Conservation*, 26(2), 4.

Kabisch, N., Frantzeskaki, N., & Hansen, R. (2022). Principles for urban nature-based solutions. *Ambio*, 51(6), 1388–1401.

Khodadad, M., Aguilar-Barajas, I., & Khan, A. Z. (2023). Green infrastructure for urban flood resilience: A review of recent literature on bibliometrics, methodologies, and typologies. *Water*, 15(3), 523.

Khorat, S., Das, D., Khatun, R., Aziz, S. M., Anand, P., Khan, A., ... & Niyogi, D. (2024). Cool roof strategies for urban thermal resilience to extreme heatwaves in tropical cities. *Energy and Buildings*, 302, 113751.

Kiss, B., Sekulova, F., Hörschelmann, K., Salk, C. F., Takahashi, W., & Wamsler, C. (2022). Citizen participation in the governance of nature-based solutions. *Environmental Policy and Governance*, 32(3), 247–272.

Knapik, E., Brandimarte, L., & Usher, M. (2024). Maintenance in sustainable stormwater management: Issues, barriers and challenges. *Journal of Environmental Planning and Management, 68*(12), 1–27.

Krishnan, S. R., Nallakaruppan, M. K., Chengoden, R., Koppu, S., Iyapparaja, M., Sadhasivam, J., & Sethuraman, S. (2022). Smart water resource management using artificial intelligence—A review. *Sustainability*, 14(20), 13384.

Kumar, P. (2021). Climate change and cities: challenges ahead. *Frontiers in Sustainable Cities*, 3, 645613.

Kumar, P., Debele, S. E., Khalili, S., Halios, C. H., Sahani, J., Aghamohammadi, N., ... & Jones, L. (2024). Urban heat mitigation by green and blue infrastructure: Drivers, effectiveness, and future needs. *The Innovation, 5*(2), 100588.

Kumar, S., & Mehany, M. S. H. M. (2022). A standardized framework for quantitative assessment of cities' socioeconomic resilience and its improvement measures. *Socio-Economic Planning Sciences*, 79, 101141.

Kumar, P., Debele, S. E., Sahani, J., Rawat, N., Marti-Cardona, B., Alfieri, S. M., ... & Zieher, T. (2021). Nature-based solutions efficiency evaluation against natural hazards:

Modelling methods, advantages and limitations. *Science of the Total Environment*, 784, 147058.

Lampinen, J., García-Antúnez, O., Olafsson, A. S., Kavanagh, K. C., Gulsrud, N. M., & Raymond, C. M. (2022). Envisioning carbon-smart and just urban green infrastructure. *Urban Forestry & Urban Greening*, 75, 127682.

Le Coent, P., Marchal, R., Hérivaux, C., Maréchal, J. C., Ladouche, B., Moncoulon, D., … & Graveline, N. (2023). Can NBS address the challenges of an urbanized Mediterranean catchment? The Lez case study. In E. López-Gunn, P. Van der Keur, N. Van Cauwenbergh, P. Le Coent, & R. Giordano (Eds.), *Greening water risks: Natural assurance schemes* (pp. 269–296). Springer International Publishing.

Leltz, E. (2022). *Nature Based Solutions for Carbon Sequestration and Biodiversity Conservation: Is there a future for solutions that look beyond tree planting for forests and grassy biomes?* (Master's thesis). Utrecht University.

Li, L., Chan, F., & Cheshmehzangi, A. (2023). Nature-based solutions and sponge city for urban water management. In F. Pacheco-torgal & C.-g. Granqvist (Eds.), *Adapting the Built Environment for Climate Change* (pp. 371–402). Woodhead Publishing. 10.1016/C2021-0-02571-2

Li, X., Liu, H., Wang, W., Zheng, Y., Lv, H., & Lv, Z. (2022). Big data analysis of the internet of things in the digital twins of smart city based on deep learning. *Future Generation Computer Systems*, 128, 167–177.

Li, X., Stringer, L. C., Chapman, S., & Dallimer, M. (2021a). How urbanization alters the intensity of the urban heat island in a tropical African city. *PLoS One*, 16(7), e0254371.

Li, L., Cheshmehzangi, A., Chan, F. K. S., & Ives, C. D. (2021b). Mapping the research landscape of nature-based solutions in urbanism. *Sustainability*, 13(7), 3876.

Liu, H. Y., Skandalos, N., Braslina, L., Kapsalis, V., & Karamanis, D. (2023, July). Integrating solar energy and nature-based solutions for climate-neutral urban environments. *Solar*, *3*(3), 382–415. https://doi.org/10.3390/solar3030022

Lv, Z., & Shang, W. (2023). Impacts of intelligent transportation systems on energy conservation and emission reduction of transport systems: A comprehensive review. *Green Technologies and Sustainability*, 1(1), 100002.

Macaione, I., Raffa, A., & Andaloro, B. (2024). Climate-Adaptive Nature-Based Regenerative Urban Green Streetscapes: Design Exploration from the City of Matera. *Sustainability*, 16(16), 6811.

Macreadie, P. I., Costa, M. D., Atwood, T. B., Friess, D. A., Kelleway, J. J., Kennedy, H., … & Duarte, C. M. (2021). Blue carbon as a natural climate solution. *Nature Reviews Earth & Environment*, 2(12), 826–839.

Mandal, M., Popek, R., Przybysz, A., Roy, A., Das, S., & Sarkar, A. (2023). Breathing fresh air in the city: implementing avenue trees as a sustainable solution to reduce particulate pollution in urban agglomerations. *Plants,* 12(7), 1545.

Marandi, A., & Main, K. L. (2021). Vulnerable City, recipient city, or climate destination? Towards a typology of domestic climate migration impacts in US cities. *Journal of environmental studies and sciences*, 11, 465–480.

Marques, A. L., Alvim, A. T. B., & Schröder, J. (2022). Ecosystem Services and Urban Planning: A Review of the Contribution of the Concept to Adaptation in Urban Areas. *Sustainability*, 14(4), 2391.

Matasov, V., Belelli Marchesini, L., Yaroslavtsev, A., Sala, G., Fareeva, O., Seregin, I., … & Valentini, R. (2020). IoT monitoring of urban tree ecosystem services: Possibilities and challenges. *Forests*, 11(7), 775.

Mavrouli, M., Mavroulis, S., Lekkas, E., & Tsakris, A. (2022). Infectious diseases associated with hydrometeorological hazards in Europe: disaster risk reduction in the context of the climate crisis and the ongoing COVID-19 pandemic. *International Journal of Environmental Research and Public Health*, 19(16), 10206.

Mihalakakou, G., Souliotis, M., Papadaki, M., Menounou, P., Dimopoulos, P., Kolokotsa, D., … & Papaefthimiou, S. (2023). Green roofs as a nature-based solution for improving urban sustainability: Progress and perspectives. *Renewable and Sustainable Energy Reviews*, 180, 113306.

Milani, A. M., & Nia, H. A. (2024). Green Infrastructure for Sustainable Urban Development: The Case of Tehran. In H. Ahmad Nia & R. Rahbarianyazd (Eds.), *Innovative approaches to cultural heritage and sustainable urban development: Integrating tradition and modernity* (pp. 90–103). Cinius Yayınları.

Mishra, A., Bajpai, S., & Bhadwal, S. (2024). Nature-Based Solutions for Climate-Resilient Development: A Technical, Policy, and Governance Perspective. In W. L. Filho, G. J. Nagy, & D. Y. Ayal (Eds.), *Handbook of Nature-Based Solutions to Mitigation and Adaptation to Climate Change* (pp. 1–24). Springer International Publishing.

Mohammed, A., Khan, A., Khan, H. S., & Santamouris, M. (2023). Cooling energy benefits of increased green infrastructure in subtropical urban building environments. *Buildings*, 13(9), 2257.

Möslinger, M., Ulpiani, G., & Vetters, N. (2023). Circular economy and waste management to empower a climate-neutral urban future. *Journal of Cleaner Production*, 421, 138454.

Murthy, M. R., Usha, T., Dash, S. K., Raju, S. K., Ramu, K., Anitha, G., ... & Ravichandran, M. (2024). Marine spatial planning for a resilient and inclusive blue economy: Lakshadweep, India, a pilot study. *Current Science*, *126*(2), 229.

Murti, R., Mathez-Stiefel, S. L., Garcia, V. R., & Rist, S. (2020). Engaging national policy makers in ecosystem-based disaster risk reduction through social learning: Lessons from workshops in Africa, oceania, Latin America and the Caribbean. *International journal of disaster risk reduction*, 45, 101463.

Naya, R. B., de la Cal Nicolás, P., Medina, C. D., Ezquerra, I., García-Pérez, S., & Monclús, J. (2023). Quality of public space and sustainable development goals: analysis of nine urban projects in Spanish cities. *Frontiers of Architectural Research*, 12(3), 477–495.

Negrello, M. (2022, June). Designing with Nature Climate-Resilient Cities: A Lesson from Copenhagen. In E. Arbizzani, E. Cangelli, C. Clemente, F. Cumo, F. Giofrè, A. M. Giovenale, M. Palme, & S. Paris (Eds.), *International Conference on Technological Imagination in the Green and Digital Transition* (pp. 853–862). Springer International Publishing.

Nita, A., Fineran, S., & Rozylowicz, L. (2022). Researchers' perspective on the main strengths and weaknesses of Environmental Impact Assessment (EIA) procedures. *Environmental Impact Assessment Review*, 92, 106690.

Nsoh, B., Katimbo, A., Guo, H., Heeren, D. M., Nakabuye, H. N., Qiao, X., … & Kiraga, S. (2024). Internet of Things-Based Automated Solutions Utilizing Machine Learning for Smart and Real-Time Irrigation Management: A Review. *Sensors (*Basel, Switzerland), *24*(23), 7480.

Nyangoko, B. P., Berg, H., Mangora, M. M., Shalli, M. S., & Gullström, M. (2022). Local perceptions of changes in mangrove ecosystem services and their implications for livelihoods and management in the Rufiji Delta, Tanzania. *Ocean & Coastal Management*, 219, 106065.

Oanh, P. T., Tamura, M., Kumano, N., & Nguyen, Q. V. (2020). Cost-benefit analysis of mixing gray and green infrastructures to adapt to sea level rise in the Vietnamese mekong river delta. *Sustainability*, 12(24), 10356.

Ojo, B. (2024). Strategies for the Optimization of Critical Infrastructure Projects to Enhance Urban Resilience to Climate Change. *The Journal of Scientific and Engineering Research*, 11, 107–123.

Oktaviani, A., & Masjud, Y. I. (2024). Nature based solution to climate change: Ecosystem based adaptation: How effective for climate change strategies? *Journal of Earth Kingdom*, 1(2), 81–90.

Olaniyi, O., Okunleye, O. J., & Olabanji, S. O. (2023). Advancing data-driven decision-making in smart cities through big data analytics: A comprehensive review of existing literature. *Current Journal of Applied Science and Technology*, 42(25), 10–18.

Osei-Kyei, R., Almeida, L. M., Ampratwum, G., & Tam, V. (2023). Systematic review of critical infrastructure resilience indicators. *Construction Innovation*, 23(5), 1210–1231.

Pan, C., Zhou, G., Shrestha, A. K., Chen, J., Kozak, R., Li, N., ... & Wang, G. (2023). Bamboo as a nature-based solution (NbS) for climate change mitigation: biomass, products, and carbon credits. *Climate*, 11(9), 175.

Pandey, B., & Ghosh, A. (2023). Urban ecosystem services and climate change: a dynamic interplay. *Frontiers in Sustainable Cities*, 5, 1281430.

Pant, D., Shah, K. K., Sharma, S., Bhatta, M., Tripathi, S., Pandey, H. P., ... & Bhat, A. K. (2023). Soil and ocean carbon sequestration, carbon capture, utilization, and storage as negative emission strategies for global climate change. *Journal of Soil Science and Plant Nutrition*, 23(2), 1421–1437.

Parihar, J., & Birman, S. (2024). Heat Resilience in Urban Environments: Strategies for Sustainable City Climate Management. In P. Singh & N. Yadav (Eds.), *The Climate-Health-Sustainability Nexus: Understanding the Interconnected Impact on Populations and the Environment* (pp. 305–324). Springer Nature Switzerland.

Paudel, P. K., Dhakal, S., & Sharma, S. (2024). Pathways of ecosystem-based disaster risk reduction: A global review of empirical evidence. *Science of The Total Environment, 929*, 172721.

Pauleit, S., Fryd, O., Backhaus, A., & Jensen, M. B. (2020). Green infrastructures to face climate change in an urbanizing world. In R. Meyers (Eds.), *Encyclopedia of sustainability science and technology* (pp. 207–234). Springer. https://doi.org/10.1007/978-1-4939-2493-6_212-3

Paxton, A. B., Riley, T. N., Steenrod, C. L., Smith, C. S., Zhang, Y. S., Gittman, R. K., ... & Davis, J. (2023). What evidence exists on the performance of nature-based solutions interventions for coastal protection in biogenic, shallow ecosystems? A systematic map protocol. *Environmental Evidence*, 12(1), 11.

Pedral Sampaio, R., Aguiar Costa, A., & Flores-Colen, I. (2022). A systematic review of artificial intelligence applied to facility management in the building information modeling context and future research directions. *Buildings*, 12(11), 1939.

Pereira, C., Flores-Colen, I., & Mendes, M. P. (2024). Guidelines to reduce the effects of urban heat in a changing climate: Green infrastructures and design measures. *Sustainable Development*, 32(1), 57–83.

Perera, C., Toxopeus, H., Klein, S., & Merfeld, K. (2024). Enabling justice for nature-based solutions in real estate development. *Nature-Based Solutions*, 6, 100148.

Petit, E. P. (2022). Smart city technologies plus nature-based solutions: viable and valuable resources for urban resilience. In J. R.vacca (Ed.), *Smart cities policies and financing* (pp. 377–398). Elsevier.

Pinto, L. V., Inácio, M., & Pereira, P. (2023). Green and blue infrastructure (GBI) and urban nature-based solutions (NbS) contribution to human and ecological well-being and health. *Oxford Open Infrastructure and Health*, 1, ouad004.

Prothi, A., Mohankumar, V., & Jha, V. (2024). Understanding riverfront developments: through time, function, and drivers for transformation. In V. R. Shinde, R. R. Mishra, U. Bhonde, & H. Vaidya (Eds.), *Managing urban rivers* (pp. 73–88). Elsevier.

Przeybilovicz, E., Cunha, M. A., Geertman, S., Leleux, C., Michels, A., Tomor, Z., ... & Meijer, A. (2022). Citizen participation in the smart city: Findings from an international comparative study. *Local Government Studies*, 48(1), 23–47.

Psistaki, K., Tsantopoulos, G., & Paschalidou, A. K. (2024). An overview of the role of forests in climate change mitigation. *Sustainability*, 16(14), 6089.

Quagliolo, C. (2023). *Spatial biophysical-economic impact assessment of Nature-Based Solutions (NBS) for urban climate change adaptation.*

Radfar, S., Mahmoudi, S., Moftakhari, H., Meckley, T., Bilskie, M. V., Collini, R., ... & Moradkhani, H. (2024). Nature-based solutions as buffers against coastal compound flooding: Exploring potential framework for process-based modeling of hazard mitigation. *Science of the Total Environment, 938*, 173529.

Raji, M., Lazarus, B. E., Nalakurthi, N. V. S. R., Tiwari, A., & Gharbia, S. (2024). Assessment of digital platforms for the quantification of co-benefits in nature-based solutions. *Sustainable Development, 33*(2), 2137–2153.

Ranjan, R. (2024). Creating synergies between payments for ecosystem services, green bonds, and catastrophe insurance markets for enhanced environmental resilience. *Land Use Policy*, 136, 106970.

Raw, J. L., Van Niekerk, L., Chauke, O., Mbatha, H., Riddin, T., & Adams, J. B. (2023). Blue carbon sinks in South Africa and the need for restoration to enhance carbon sequestration. *Science of the Total Environment*, 859, 160142.

Riffat, S., Ahmad, M. I., & Shakir, A. (2024). Eco-cities: Sustainable urban living. In S. Riffat, M. I. Ahmad, & A. Shakir (Eds.), *Sustainable energy technologies and low carbon buildings* (pp. 259–309). Springer Nature Switzerland.

Rosales, R. C., Bellino, P., Elnahass, M., Heubaum, H., Lim, P., Lemaistre, P., ... & Sjögersten, S. (2021). *Voluntary carbon markets in ASEAN: Challenges and opportunities for scaling up*. UK Government/The British High Commission Singapore. https://www.gov.uk/government/publications/uk-singapore-cop26-universities-network-policy-reports

Scolobig, A., Rodriguez, J. A., Martin, J., & Linnerooth-Bayer, J. (2023). *Governance innovation for the design, financing and implementation of NBS, and their application to the concept and demonstration projects.*

Seddon, N., Smith, A., Smith, P., Key, I., Chausson, A., Girardin, C., ... & Turner, B. (2021). Getting the message right on nature-based solutions to climate change. *Global Change Biology*, 27(8), 1518–1546.

Serdar, M. Z., Koç, M., & Al-Ghamdi, S. G. (2022). Urban transportation networks resilience: indicators, disturbances, and assessment methods. *Sustainable Cities and Society*, 76, 103452.

Shao, H., & Kim, G. (2022). A comprehensive review of different types of green infrastructure to mitigate urban heat islands: Progress, functions, and benefits. *Land*, 11(10), 1792.

Sharifi, A., Beris, A. T., Javidi, A. S., Nouri, M. S., Lonbar, A. G., & Ahmadi, M. (2024). Application of artificial intelligence in digital twin models for stormwater infrastructure systems in smart cities. *Advanced Engineering Informatics*, 61, 102485.

Silva Martins da Cunha Marinho, M. G., dos Santos Batista, R., Porto, P., Nemi, A., & de Campos, C. (2024). The modernization of medical education in Brazil: Rockefeller foundation funding and the Ribeirão Preto medical school in a development context (1951–1964). *Historia Crítica, 93*, 53–78.

Song, C. (2022). Application of nature-based measures in China's sponge city initiative: Current trends and perspectives. *Nature-Based Solutions*, 2, 100010.

Sonkar, G. (2024). Augmenting carbon sink capacity through nature based solutions for climate change mitigation: A case study of India. In P. Singh, P. Srivastava, & A. Sorokin (Eds.), *Nature-based solutions in achieving sustainable development goals: Harmonizing nature and progress* (pp. 207–231). Springer Nature Switzerland.

Sousa-Silva, R., Duflos, M., Barona, C. O., & Paquette, A. (2023). Keys to better planning and integrating urban tree planting initiatives. *Landscape and Urban Planning*, 231, 104649.

Sowińska-Świerkosz, B., & García, J. (2022). What are nature-based solutions (NBS)? Setting core ideas for concept clarification. *Nature-Based Solutions*, 2, 100009.

Strain, E. M. A., Kompas, T., Boxshall, A., Kelvin, J., Swearer, S., & Morris, R. L. (2022). Assessing the coastal protection services of natural mangrove forests and artificial rock revetments. *Ecosystem Services*, 55, 101429.

Sudarma, I., Saifulloh, M., Diara, I. W., & As-Syakur, A. (2024). Carbon stocks dynamics of urban green space ecosystems using time-series vegetation indices. *Ecological Engineering & Environmental Technology*, 25, 147–162.

Suel, E., Lynch, C., Wood, M., Murat, T., Casey, G., & Dennett, A. (2024). Measuring transport-associated urban inequalities: Where are we and where do we go from here? *Transport Reviews*, *44*(6), 1235–1257.

Sukanya, R., & Tantia, V. (2023). Urbanization and the impact on economic development. In J. Martínez-Falcó, B. Marco-Lajara, E. Sánchez-García, & L. Millan-Tudela (Eds.), *New perspectives and possibilities in strategic management in the 21st century: Between tradition and modernity* (pp. 369–408). IGI Global Scientific Publishing.

Sunkur, R., Kantamaneni, K., Bokhoree, C., & Ravan, S. (2023). Mangroves' role in supporting ecosystem-based techniques to reduce disaster risk and adapt to climate change: A review. *Journal of Sea Research*, *196*, 102449.

Susca, T. (2022). Nature-based solutions applied to the built environment to alleviate climate change: Benefits, co-benefits, and trade-offs in a geographical multi-scale perspective. In M. Lackner, B. Sajjadi, & W.-Y. Chen (Eds.), *Handbook of climate change mitigation and adaptation* (pp. 2117–2167). Springer International Publishing.

Tainio, M., Andersen, Z. J., Nieuwenhuijsen, M. J., Hu, L., De Nazelle, A., An, R., ... & de Sá, T. H. (2021). Air pollution, physical activity and health: A mapping review of the evidence. *Environment international*, 147, 105954.

Tan, T., Kong, F., Yin, H., Cook, L. M., Middel, A., & Yang, S. (2023). Carbon dioxide reduction from green roofs: A comprehensive review of processes, factors, and quantitative methods. *Renewable and Sustainable Energy Reviews*, 182, 113412.

Terdiman, M., & Angert, T. (2023). Green and smart cities in the developing world. In R. C. Brears (Ed.), *The Palgrave encyclopedia of urban and regional futures* (pp. 676–684). Springer International Publishing.

Thomas, V. (2023). *Risk and resilience in the era of climate change*. Palgrave Macmillan.

Thompson, A., Bunds, K., Larson, L., Cutts, B., & Hipp, J. A. (2023). Paying for nature-based solutions: A review of funding and financing mechanisms for ecosystem services and their impacts on social equity. *Sustainable Development*, 31(4), 1991–2066.

Tomar, J. M. S., Kumar, R., Kaushal, R., Kar, S. K., Mehta, H., & Chaturvedi, O. P. (2023). Soil conservation measures improve vegetation development and ecological processes in the Himalayan slopes. *Tropical Ecology*, 64(1), 53–61.

Tozer, L., Bulkeley, H., & Xie, L. (2022a). Transnational governance and the urban politics of nature-based solutions for climate change. *Global Environmental Politics*, 22(3), 81–103.

Tozer, L., Bulkeley, H., van der Jagt, A., Toxopeus, H., Xie, L., & Runhaar, H. (2022b). Catalyzing sustainability pathways: Navigating urban nature-based solutions in Europe. *Global Environmental Change*, 74, 102521.

Tun, K. Z., Pramanik, M., Chakrabortty, R., Chowdhury, K., Halder, B., Pande, C. B., ... & Zhran, M. (2024). Mainstreaming nature-based solutions for climate adaptation in southeast Asia: A systematic review. *Earth Systems and Environment*, *9*(2), 1333–1351.

Ulpiani, G., Rebolledo, E., Vetters, N., Florio, P., & Bertoldi, P. (2023). Funding and financing the zero emissions journey: Urban visions from the 100 climate-neutral and smart cities mission. *Humanities and Social Sciences Communications*, 10(1), 1–14.

Umoh, A. A., Ohenhen, P. E., Chidolue, O., Ngozichukwu, B., Fafure, A. V., & Ibekwe, K. I. (2024). Incorporating energy efficiency in urban planning: A review of policies and best practices. *Engineering Science & Technology Journal*, 5(1), 83–98.

Verma, A., Schmidt-Vogt, D., De Alban, J. D. T., Lim, C. L., & Webb, E. L. (2021). Drivers and mechanisms of forest change in the Himalayas. *Global Environmental Change*, 68, 102244.

Vikas, N., & Hari, S. (2023). Innovative financing mechanisms to leverage ecosystem based adaptation (EbA) finance for vulnerable communities. *Global Economic Observer*, 11(1), 116–125.

Vögt, V., Harrs, J. A., Reinhart, V., Hollenbach, P., Bühler, M. M., & Tewes, T. (2023). Implementing agile data workflows to unlock climate-resilient urban planning. *Climate*, 11(9), 174.

Wang, J., Xiang, Z., Wang, W., Chang, W., & Wang, Y. (2021). Impacts of strengthened warming by urban heat island on carbon sequestration of urban ecosystems in a subtropical city of China. *Urban Ecosystems*, 24, 1165–1177.

Wang, M., Zhong, X., Yuan, H., Zhang, D., Cheng, L., & Zhang, J. (2024). Integrating carbon and water footprint into nature-based solution (NBS) for urban planning in a highly built-up area in Guangzhou, China. *Science of the Total Environment*, 951, 175505.

Wild, T., Baptista, M., Wilker, J., Kanai, J. M., Giusti, M., Henderson, H., … & Kozak, D. (2024). Valuation of urban nature-based solutions in Latin American and European cities. *Urban Forestry & Urban Greening*, 91, 128162.

Wong, N. H., Tan, C. L., Kolokotsa, D. D., & Takebayashi, H. (2021). Greenery as a mitigation and adaptation strategy to urban heat. *Nature Reviews Earth & Environment*, 2(3), 166–181.

Wudu, K., Abegaz, A., Ayele, L., & Ybabe, M. (2023). The impacts of climate change on biodiversity loss and its remedial measures using nature-based conservation approach: A global perspective. *Biodiversity and Conservation*, 32(12), 3681–3701.

Xiao, Y., Li, C., Li, X., Wang, Y., Sun, P., Xu, X., … & Xu, W. (2024). International mangrove carbon sink research analysis. *Regional Studies in Marine Science*, *77*, 103681.

Xiao, Z., Ge, H., Lacasse, M. A., Wang, L., & Zmeureanu, R. (2023). Nature-based solutions for carbon neutral climate resilient buildings and communities: A review of technical evidence, design guidelines, and policies. *Buildings*, 13(6), 1389.

Xie, W., Sun, C., & Lin, Z. (2023). Spatial-temporal evolution of urban form resilience to climate disturbance in adaptive cycle: A case study of Changchun city. *Urban Climate*, 49, 101461.

Yadav, N., Rakholia, S., & Yosef, R. (2024). Decision support systems in forestry and tree-planting practices and the prioritization of ecosystem services: A review. *Land*, 13(2), 230.

Zhao, D., Cai, J., Xu, Y., Liu, Y., & Yao, M. (2023). Carbon sinks in urban public green spaces under carbon neutrality: A bibliometric analysis and systematic literature review. *Urban Forestry & Urban Greening*, *86*, 128037.

Zhou, Y. (2023). Worldwide carbon neutrality transition? Energy efficiency, renewable, carbon trading and advanced energy policies. *Energy Reviews*, 2(2), 100026.

Zhu, T., Li, C., Liu, X., & Zhao, X. (2024). Assessing ecological restoration in arid mining regions: A progressive evaluation system. *Sustainability*, 16(6), 2266.

Section III

Tools and Techniques to Explore Urban Form-Energy Use-Carbon Emission Link

5 Water Management in Climate-Related Disasters and Urban Resilience

Chukwuemeka Kingsley John,
Fidelis Odedishemi Ajibade, Pankaj Kumar,
Temitope Fausat Ajibade,
Kayode Hassan Lasisi, Bashir Adelodun,
Abdulhamid Yusuf, and James Rotimi Adewumi

5.1 INTRODUCTION

Water management is a critical component of urban sustainability, especially as cities grapple with the increasing frequency and severity of climate-related disasters. Urban water systems are directly impacted by the problems caused by climate change, which include longer droughts, more powerful floods, and unpredictable weather patterns that jeopardize the quantity and quality of available water resources. The effects of these issues are magnified in highly populated urban regions, increasing the dangers to public health, economic stability, and infrastructural resilience. In this situation, effective water management must not only address short-term risks associated with disasters but also work towards long-term sustainability objectives, such as achieving carbon neutrality (Muyambo et al., 2023; du Plessis, 2023). To minimize their environmental impact and enable urban areas to adapt to changing climate conditions, integration is essential. It is becoming more widely acknowledged that creating sustainable cities requires addressing the relationship between water management and urban resilience. The ability of cities to withstand shocks, strains, and adaptations, including those made worse by climate change, is referred to as urban resilience. Since water supports many facets of urban life, including agriculture, industry, sanitation, and the supply of drinking water, it is a crucial resource in this equation. But there are serious problems because of how unpredictable water availability is becoming because of climate change (Debangshi et al., 2022; Khalil et al., 2022). Therefore, cities need to create strong water management plans that increase their ability to withstand climate-related calamities. These tactics, which include protecting water quality, preparing for droughts, and managing floods, must all be included in larger urban planning initiatives. Water management needs to be

DOI: 10.1201/9781003518037-8

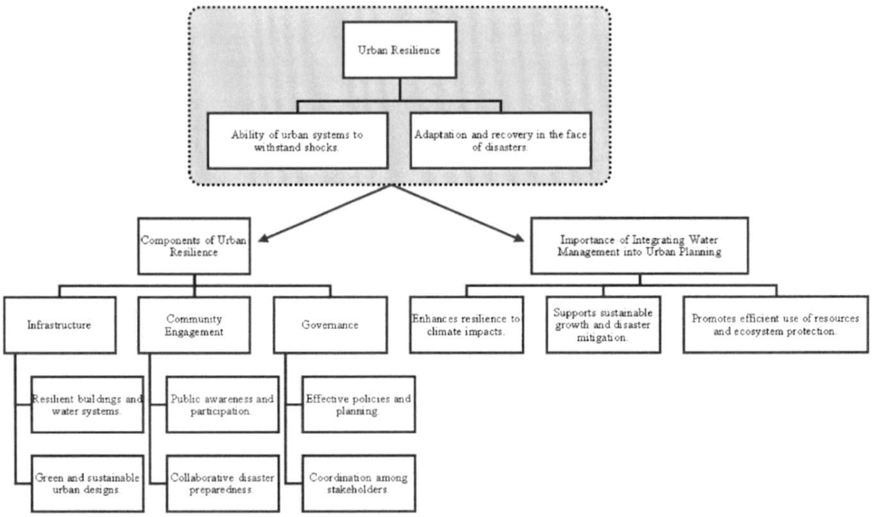

FIGURE 5.1 Framework for urban resilience, outlining its definition, key components, and the role of integrating water management into urban planning.

considered from a carbon-neutral perspective in the context of urban sustainability. It is crucial that water management techniques support rather than undermine efforts by cities to lower their carbon footprints. To lower the energy demand related to water treatment and distribution, this calls for a change towards more energy-efficient water infrastructure, the use of naturally occurring solutions that trap carbon, and the encouragement of water conservation (Bonato et al., 2022; Kılkış et al., 2023). Cities can simultaneously address the difficulties of climate-related disasters and improve their overall sustainability by coordinating water management with carbon neutrality aspirations. Building resilient, sustainable urban landscapes that can survive the challenges of a changing climate requires an integrated strategy (Figure 5.1).

Building resilient cities that can endure the effects of climate-related calamities and move towards sustainability goals, such as carbon neutrality, requires the integration of water management into urban planning. Urban regions are more susceptible than other places to water-related problems like drought, flooding, and stormwater runoff, all of which are made worse by global warming. Water management must therefore be considered in effective urban planning, not as a separate problem but as a key element of city development. Through this integration, communities are protected from the growing hazards posed by a changing climate, infrastructure is built to withstand extreme weather events, and water resources are managed sustainably. Furthermore, cities may optimize land use, improve green areas, and implement nature-based solutions that not only mitigate climate impacts but also aid in carbon sequestration by incorporating water management into urban planning (Dąbrowska et al., 2023; Mohammadpourlima et al., 2024; Piemontese et al., 2024). Beyond being immediately ready for emergencies, water management is crucial to urban development. It is essential to reach carbon neutrality and long-term urban sustainability. Conventional urban planning techniques frequently fail to consider the links between energy use, carbon emissions, and water systems, which results in inefficiencies and lost chances for positive effects on the environment and the economy. Cities may lessen their carbon footprint through energy-efficient water systems, encourage water reuse and recycling, and limit the need for carbon-intensive infrastructure repairs following disasters by including water management in the planning process. This holistic approach not only enhances urban resilience but also supports the global transition to a low-carbon economy, making cities more sustainable and liveable for future generations (de Oliveira et al., 2022; Chen et al., 2023; Mohammadpourlima et al., 2024).

Though its direct contributions to carbon reduction require more precise explanations, water management is essential to reaching carbon neutrality. Energy-intensive procedures that greatly increase greenhouse gas emissions are urban water systems, which include supply, distribution, and wastewater treatment. Cities may minimize their overall carbon footprint by integrating energy-efficient water infrastructure and reducing their need for fossil fuel-based energy. Modern wastewater treatment techniques can produce biogas for use in renewable energy sources, while leak detection and pressure control can optimize water distribution networks and lower the energy needed for pumping. These tactics support wider climate action objectives while also improving urban resilience (Zhang et al., 2021; Wan et al., 2022; Abulibdeh, 2024). A thorough investigation of the deployment of energy-efficient water systems is necessary to bolster the case even more. Technologies that can drastically cut energy use include solar-powered water treatment facilities, low-energy desalination, and gravity-fed water supplies. The requirement for vast pipeline networks is further reduced by decentralized water treatment facilities, which lowers emissions and energy consumption. Using automation and real-time monitoring, smart water management systems also minimize wasteful energy usage and maximize resource utilization. These developments show how water security may be preserved while lowering urban carbon emissions through effective water management (Huang et al., 2023; Bhardwaj et al., 2024; Manimaran, 2025). It is important to highlight how these strategies affect urban carbon footprints. Cities can reduce emissions from electricity generation, which is frequently the biggest contributor to urban carbon footprints, by lowering the energy demand in water management. The incorporation of energy and water efficiency measures into urban planning also promotes a comprehensive approach to sustainability by encouraging synergies with green building designs and the use of renewable energy. Accelerating the transition to carbon neutrality will require regulations and investments in energy-efficient water infrastructure. Thus, a more thorough description of these tactics' operations and observable advantages will improve the conversation about water management's contribution to carbon reduction (Shadbahr et al., 2022; Hui et al., 2023).

The aim of the chapter is to explore the critical role of water management in enhancing urban resilience against climate-related disasters while contributing to broader sustainability and carbon neutrality goals. The objective of this chapter is to give readers a thorough grasp of how efficient water management techniques can lessen the effects of droughts, floods, and other climate-related problems in urban areas. Analysing how water management is incorporated into urban design, showcasing cutting-edge strategies that support carbon reduction goals, and stressing the significance of technology, community involvement, and governance in building resilient, sustainable cities are among the goals. This chapter aims to provide insights and useful advice for policymakers, urban planners, and stakeholders to foster urban environments that are both resilient to climate change and committed to reducing their environmental footprint by tying water management with the pursuit of carbon neutrality.

5.2 RESULTS AND DISCUSSION

5.2.1 Impact of Climate-Related Disasters on Water Resources

Climate-related disasters have profound impacts on water resources, affecting both supply and quality. Floods have the potential to destroy water infrastructure, seriously harm pipelines and treatment plants, contaminate sources of drinking water, and raise the risk of waterborne illnesses (Figure 5.2). Conversely, droughts weaken the availability of water, put pressure on supply chains, and intensify rivalry for scarce resources, which lowers water quality and increases urban vulnerability.

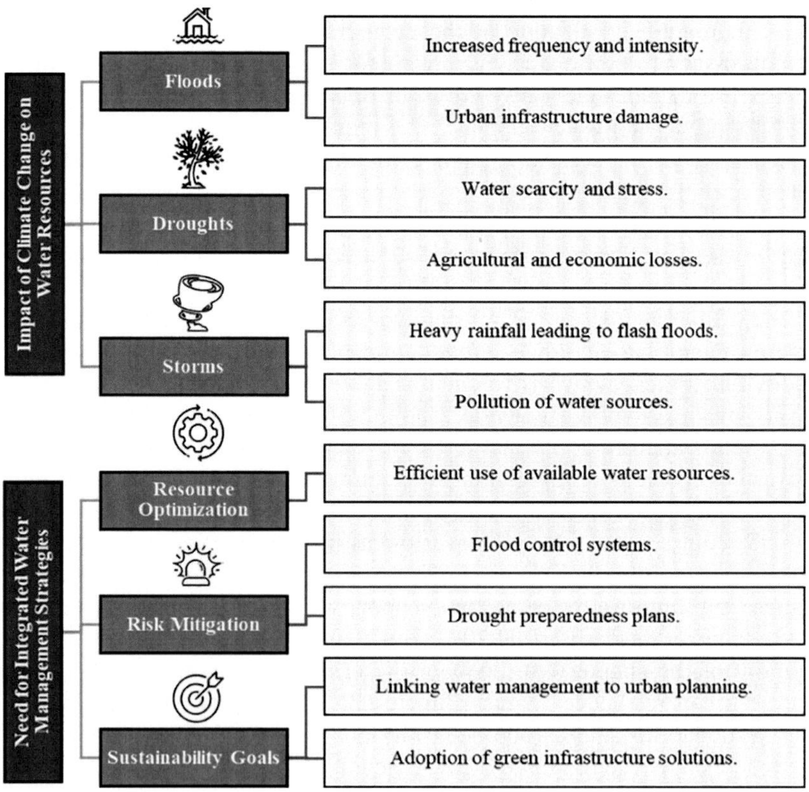

FIGURE 5.2 Framework depicting the impacts of climate change on water resources and the necessity of integrated management strategies.

Storms have the potential to worsen these problems by directly damaging infrastructure as well as indirectly by eroding soil and creating runoff that lowers water quality. To lessen the negative effects of climate-related disasters on urban water systems and guarantee a steady, safe water supply, these consequences highlight the necessity for robust water infrastructure and adaptive management solutions (Crawford et al., 2022; Richardson et al., 2024).

i. Effects of floods, droughts, and storms on water supply and quality
   Storms, droughts, and floods all have a major impact on the quantity and quality of urban water supplies, which makes cities far less resilient and sustainable. For example, floods have the potential to overrun water infrastructure and contaminate water sources with harmful substances, germs, and contaminants. Due to this contamination, drinking water sources may become dangerous, posing a risk to public health and requiring expensive cleanup procedures. Furthermore, floods frequently destroy vital infrastructure, including distribution networks and water treatment facilities, interrupting the flow of clean water and raising the danger of waterborne illnesses. When sediments, nutrients, and chemicals are pushed into water bodies following flooding, it can also result in longer term degradation of water quality, making attempts to manage and restore urban water resources even more difficult (Crawford et al., 2022; Pokhrel et al., 2022; Mukhopadhyay et al., 2022). Conversely, droughts put a burden on the water supply since they make freshwater supplies less available, which frequently results in excessive groundwater extraction and reservoir depletion. This scarcity exacerbates problems with water quality by concentrating pollutants in water sources and affecting the immediate availability of water for drinking, sanitation, and industrial use. Long-term droughts can cause water supplies to become more salinized, especially in coastal areas, rendering the water unfit for many uses without the need for costly desalination procedures. Storms that provide sudden, heavy runoff that overwhelms drainage systems and introduces pollutants into rivers, lakes, and groundwater sources, such as hurricanes and typhoons, can worsen the state of the water supply (Singh et al., 2022; Mishra, 2023; Richardson et al., 2024). Urban water management faces a complex and interconnected set of issues due to the combined consequences of these climate-related calamities. These challenges call for integrated measures that improve sustainability and resilience.

   The carbon footprint of conventional water management systems is high because they depend on energy-intensive processes including water extraction, treatment, distribution, and wastewater management. Frequently relying on energy sources driven by fossil fuels, these systems contribute to greenhouse gas emissions. The carbon footprint of metropolitan infrastructure is increased by the need for significant electricity for long-distance water pumping and traditional wastewater treatment techniques. Energy use and emissions are also made worse by ageing infrastructure that is inefficient, including leaks and antiquated treatment technology (Grzegorzek et al., 2023; Chen et al., 2024; Hauashdh et al., 2024). Alternatives for sustainable water management provide workable ways to cut these emissions. Energy consumption can be greatly reduced by using energy-efficient technologies, including anaerobic digestion in wastewater treatment, gravity-fed distribution systems, and sophisticated membrane filtration. Reliance on fossil fuels is further decreased by integrating renewable energy sources, such as solar-powered desalination and biogas recovery from sewage sludge. Furthermore, decentralized water systems, such as localized wastewater treatment and rainwater collection, reduce the need for large, energy-intensive infrastructure, which lowers carbon footprints overall (Okeke et al., 2022). Cities and industries can reduce emissions while enhancing water security and resilience by switching to these sustainable options. Urban areas' carbon footprint is directly reduced when energy demand for water management is reduced, aiding in the fight against climate change. The transition to sustainability can be accelerated by putting laws into place that support water conservation, infrastructure modernization, and the use of renewable energy in water systems. Thus, a better understanding of how water management contributes to carbon neutrality can be obtained by contrasting the advantages of sustainable options with the carbon footprint of conventional systems (Raihan et al., 2022; Ssekyanzi et al., 2024).

ii. Implications for urban areas and their water infrastructure

There are substantial hazards to the operation of cities and the welfare of their citizens associated with climate-related disasters, which have far-reaching effects on metropolitan areas and their water infrastructure. For example, extensive damage to water infrastructure, such as distribution networks, pumping stations, and treatment facilities, can result from flooding (Figure 5.2). It can be challenging to maintain public health and hygiene during and after a disaster when these vital facilities are submerged since it might cause rapid disruptions to the water supply and sanitation services. Infrastructure degradation or destruction often leads to expensive repairs and protracted service interruptions, which disproportionately impact people who are already vulnerable (Ridha et al., 2022; Hakovirta, 2024). In addition, the problem is made more difficult by the tainted water supplies that result from floods. This is because metropolitan regions may be forced to use unsustainable emergency methods like water trucking or bottled water distribution. Storms and droughts make urban water infrastructure even less resilient. Long-lasting droughts make it harder for communities to access surface and groundwater resources, forcing them to draw too much from their current supplies or spend money on more expensive and energy-intensive alternatives like desalination or water recycling. Droughts also put additional strain on the infrastructure already in place, which increases the risk of water delivery systems failing due to their inability to satisfy demand (Zozmann et al., 2022; Ayaz et al., 2022). Storms can cause physical damage to infrastructure, such as pipeline collapses or riverbank erosion close to vital facilities, especially when they combine strong winds and precipitation. Furthermore, a rapid surge in rainwater can overwhelm combined sewer systems, spilling raw sewage into streams and worsening the situation for the general public's health. These difficulties highlight the necessity for communities to make investments in adaptable, resilient water infrastructure that can both serve larger sustainability and carbon neutrality goals and endure and recover from the effects of climate-related calamities (Farooq et al., 2023; Owolabi et al., 2022; Lebu et al., 2024).

5.2.2 Urban Resilience and Water Management

i. Definition and components of urban resilience
The ability of a city to withstand, bounce back from, and adjust to a variety of shocks and strains, including those brought on by climate-related disasters is referred to as urban resilience. Urban regions need to be resilient to preserve public health, keep up their functionality, and prosper in the face of growing environmental and social issues. Urban resilience is a notion that includes the ability to adapt to long-term changes in climate patterns, economic situations, and demographic transitions, in addition to the ability to absorb immediate shocks like floods, droughts, and storms. Urban resilience is a comprehensive strategy that incorporates social, economic, environmental, and physical systems to make sure that cities are not only ready for future calamities but can also adapt and grow over time to meet new problems. Urban resilience is made up of several different structural and non-structural components (Debangshi et al., 2022; Amirzadeh et al., 2022; Zeng et al., 2022). Resilient infrastructure is made up of stormwater management systems, water treatment plants, and flood barriers that are built to resist extreme weather events and quickly recover from them. These elements also incorporate the use of green infrastructure, which can lessen urban heat islands, absorb surplus rainfall, and give ecological advantages. Examples of this infrastructure include wetlands and green roofs. Non-structural elements, which are equally significant, include community engagement tactics, thorough disaster preparedness plans, and efficient governance structures. These elements guarantee that cities can effectively distribute resources, plan responses, and encourage a resilient culture among their citizens. To predict and address new hazards, urban resilience also necessitates ongoing data collection, analysis, and monitoring (Kumar et al., 2024a; Ojo, 2024). Policies that support fairness and sustainability are also necessary to guarantee that all segments of the urban population are safe and able to recover from disasters. These elements work together to provide a strong framework that helps cities deal with the difficulties posed by climate change and grow as well.

ii. Role of water management in enhancing urban resilience
Water management plays a pivotal role in enhancing urban resilience, particularly in the face of climate-related disasters. Climate change is making storms, droughts, and floods more frequent and severe; thus, it is imperative to implement effective water management methods to lessen their effects. Cities can enhance their ability to regulate and allocate water resources to

avert floods, sustain water supplies during dry spells, and safeguard water quality from pollution under severe weather conditions by putting integrated water management systems into place. To lessen the risk of flooding in metropolitan areas, for example, flood defences including levees, retention basins, and stormwater drainage systems are built to help manage surplus water during heavy rainfall (Garcia et al., 2022; Rezvani et al., 2023). In a similar vein, water conservation efforts, the advancement of water-efficient technologies, and the creation of substitute water sources, including wastewater recycling and rainwater collecting, guarantee that cities are resilient to water scarcity during droughts. These solutions provide a consistent supply of clean water, even in times of emergency, protecting not just urban infrastructure but also the health and well-being of urban inhabitants. Water management is essential for the long-term resilience of metropolitan communities, even beyond the immediate reaction to disasters. Cities may develop more adaptable and flexible systems that react to changing climate conditions by including water management in larger urban planning and sustainability projects. For instance, using green infrastructure—permeable pavements, green roofs, and restored wetlands—aligns water management with carbon neutrality objectives by improving water absorption, lowering runoff, and contributing to carbon sequestration. Furthermore, by using real-time data and predictive analytics, smart water management systems can enhance the efficacy and efficiency of urban water consumption, assisting cities in anticipating and lessening the effects of climate-related catastrophes before they happen (Lee et al., 2022a; La Vigna, 2022; Pellegrino & Ahern, 2023). These methods help create a more resilient urban environment that can withstand short-term shocks as well as long-term strains, guaranteeing that water management promotes cities' overall sustainability and climate change adaptation.

Green infrastructure, which incorporates natural systems into urban settings, is essential to achieving carbon neutrality goals. By improving carbon sequestration, elements like wetlands, urban forests, green roofs, and permeable pavements aid in reducing the effects of climate change. In addition to enhancing air quality and lowering the urban heat island effect, vegetation in green infrastructure absorbs carbon dioxide from the atmosphere, lowering the need for energy for cooling. Additionally, by naturally filtering contaminants and controlling runoff, these natural solutions lessen dependency on energy-intensive water treatment and distribution facilities. Urban areas can successfully lower emissions and improve environmental resilience by integrating green spaces into city planning (Woon et al., 2023; Aghaloo et al., 2024; Azmeer et al., 2024). Decentralized water treatment and stormwater management lower energy consumption and greenhouse gas emissions, assisting in the attainment of carbon neutrality. Transporting, treating, and disposing of water requires a significant amount of infrastructure and energy in conventional water management systems. However, decentralized techniques such as on-site greywater recycling, rainfall collecting, and localized wastewater treatment significantly reduce the need for energy-intensive pumping and processing. Similarly, bioswales and constructed wetlands are examples of sustainable stormwater solutions that naturally manage runoff, reducing the need for mechanical treatment and preventing overburdened drainage systems. By combining these tactics, cities can lower their carbon footprint, improve water efficiency, and move closer to becoming carbon neutral (Lam et al., 2022; Lakho et al., 2022; John & Pu, 2023).

5.2.3 Strategies for Water Management in Climate-Related Disasters

Water management options that work in climate-related disasters include a variety of approaches that are intended to mitigate floods and droughts while maintaining a consistent emergency water supply (Figure 5.3). The goal of flood management strategies, which include building levees and zoning regions within floodplains, is to shield susceptible areas from excessive precipitation and increasing water levels. On the other hand, drought management tactics emphasize the use of rainwater collection and improved water conservation to guarantee a steady supply of water during times of shortage. Furthermore, in the wake of disasters, having enough emergency water supply planning and infrastructure is essential for ensuring that people have access to clean water right away. Cities may establish a comprehensive framework for managing water resources that not only lessens the effects of extreme weather events but also guarantees resilience and flexibility in the face of persistent climate change by using these measures (Chaurasia et al., 2022; Bibi & Kara, 2023).

i. Flood management techniques (e.g., levees, floodplain zoning)
   Flood management techniques are essential for protecting urban areas from the devastating effects of floods, which are becoming more frequent and severe due to climate change. Building levees—raised embankments—is one of the most popular methods used to stop river water from overflowing into nearby communities during times of high flow. By acting as barriers, levees direct water away from populous or dangerous regions and follow its natural route. Levees are useful, but they need to be carefully maintained

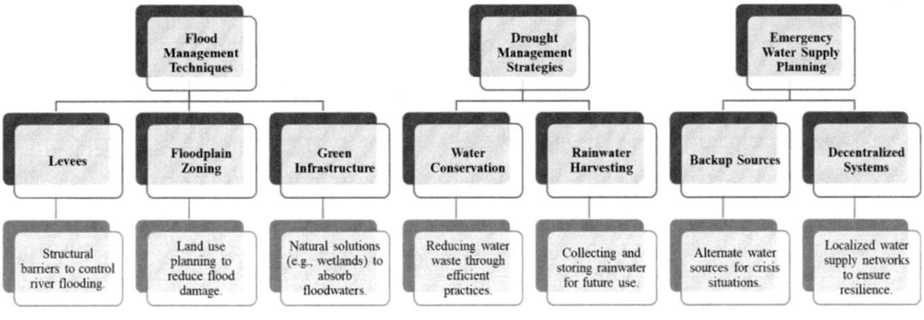

FIGURE 5.3 Key strategies for water management in climate-related disasters, highlighting flood management techniques, drought mitigation approaches, and emergency water supply planning.

and inspected to guarantee their integrity because breaches can cause disastrous flooding. Floodplain zoning is an essential flood management technique in addition to levees. This method controls how land is used in flood-prone locations, frequently by prohibiting development or mandating higher, flood-resistant structures. Cities can lessen the possibility of damage and guarantee that floodwaters can safely spread out without affecting vital infrastructure or residential areas by designating specific regions as floodplains and restricting their use (Junger et al., 2022; Serra-Llobet et al., 2022; Bibi & Kara, 2023). Floodplain zoning preserves natural floodplains and wetlands that function as natural buffers, which improves the resilience of urban environments in addition to being a preventive strategy. These places can absorb extra water, which lessens the force and speed of floodwaters downstream. By encouraging the preservation of natural landscapes and reducing the need for substantial constructed solutions, this strategy is consistent with sustainable urban development. Furthermore, to establish a multi-layered defence system, modern flood control frequently combines conventional constructed solutions—like levees—with natural solutions—like restoring floodplains and creating urban green spaces (Huang et al., 2022; Knuijt, 2022; Vitale, 2023). This integrated approach preserves ecological services and improves the city's ability to adapt to changing climate conditions, protecting urban areas from immediate flood hazards while also promoting long-term resilience.

ii. Drought management strategies (e.g., water conservation, rainwater harvesting)

Drought management strategies are critical for ensuring a reliable water supply in urban areas during periods of water scarcity. Water conservation is one of the best tactics; it entails using less water through a variety of techniques, such as public awareness campaigns, the adoption of water-efficient technologies, and the enforcement of laws that restrict the use of water for non-essential purposes. Cities should, for instance, promote or require the use of drought-resistant landscaping, low-flow plumbing, and effective irrigation systems to reduce the amount of water used in residences, commercial buildings, and public areas. Campaigns for public education are essential in creating a culture of conservation because they motivate people to adopt habits like leak repair, water-saving appliance use, and water waste reduction. Cities may greatly lessen the demand for their current water resources and increase their resilience to drought situations by promoting water conservation (Chaurasia et al., 2022; Singha et al., 2022; Kumar et al., 2024b). Rainwater harvesting, or collecting and storing rainwater for later use, is another essential drought control technique. This can be accomplished in several ways, from straightforward rain buckets connected to downspouts to more intricate systems that gather and store substantial amounts of water for use in industry, irrigation, or even for human consumption after appropriate treatment. In addition to offering a backup water supply during dry spells, rainwater collection eases the strain on municipal water systems and replenishes groundwater reserves. Rainwater harvesting can also help control stormwater runoff in urban settings where impermeable surfaces like roads and buildings hinder natural water infiltration, lowering the danger of flooding while concurrently addressing water scarcity. Rainwater harvesting can be incorporated into building standards, urban planning, and community behaviours to help cities become more drought-resistant and provide a more sustainable and dependable water supply in the face of climate change (John et al., 2021; Roba et al., 2022; John & Pu, 2023).

iii. Emergency water supply planning and infrastructure
Urban resilience depends on emergency water supply planning and infrastructure, which guarantees that cities can supply enough clean water to their citizens in the event of a climate calamity. A key component of emergency water supply planning is creating backup plans that account for possible water supply interruptions brought on by events like droughts, floods, or infrastructure failures. These plans usually involve the identification of backup water sources that can be used if regular supplies are disrupted, such as reservoirs, groundwater wells, or emergency desalination plants. Cities may also set up makeshift pipes or mobile water treatment units as emergency distribution networks to supply water to impacted areas. Cities can minimize the impact on public health and sustain key services during emergencies by putting these plans and processes in place and ensuring a quick response (Krueger et al., 2022; Council et al., 2022). Not less significant is the infrastructure needed to supply emergency water, which frequently consists of buildings and machinery built with disaster recovery in mind. For instance, water treatment facilities and pumping stations depend on backup power systems to continue operating when there is a power outage due to a storm or other occurrence. Additionally, cities can spend money on decentralized water treatment facilities, which can function both independently and in conjunction with the main system to provide redundancy if the primary infrastructure fails. In addition, in the case of an interruption, emergency water storage tanks placed thoughtfully across cities can supply clean water right away. To make sure they are prepared for use when needed, these tanks are normally filled and maintained regularly (Rodriguez-Garcia et al., 2022; Wutich et al., 2023). Cities may improve their ability to withstand climate-related disasters by making significant investments in emergency water supply infrastructure and planning. This will guarantee that water will always be available and safe, even in the most difficult situations.

5.2.4 Case Studies

Case studies demonstrating the effective implementation of water management solutions in urban and regional settings are presented in this section, highlighting their practical applicability and contribution to increased climate resilience. Through an analysis of instances such as Singapore's varied water supply approach and Copenhagen's inventive Cloudburst Management Plan, we may observe how these cities have used cutting-edge technologies and comprehensive methods to tackle water management issues. These case studies offer important insights into the lessons discovered from previous climate-related disasters in addition to highlighting successful practices. Resilience and adaptation are crucial, as demonstrated by the progress Japan made in water infrastructure after the Great East Japan Earthquake and the shortcomings of New Orleans' flood management systems before and after Hurricane Katrina. By combining these effective tactics with the knowledge gained from previous occurrences, cities may create resilient, flexible water management systems that support urban sustainability and carbon neutrality and better equip them to handle climate-related issues in the future (van der Berg, 2023; Gulsrud et al., 2023; Khaspuria et al., 2024).

i. Examples of cities or regions that have successfully implemented water management strategies for climate resilience
   Globally, cities have exhibited inventive methods for managing water resources, which augment climate adaptability and sustainability. For example, the Danish city of Copenhagen has made great progress in incorporating climate resilience into its water management plan. The Cloudburst Management Plan, a comprehensive plan that the city has put into effect, focuses on using green infrastructure to manage excessive rainfall. Copenhagen's Cloudburst Management Plan incorporates green infrastructure, like permeable surfaces and urban green spaces, which reduce the risk of flooding and help achieve carbon neutrality by sequestering carbon. Energy-intensive water treatment requirements are decreased, emissions are decreased, and the city is in line with its climate goals by giving priority to nature-based solutions and decentralized stormwater management. Copenhagen has efficiently decreased flood hazards and increased urban green spaces by incorporating water retention systems into urban design and by establishing vast, multipurpose parks. In addition to managing stormwater, this strategy improves the city's liveability and climate resilience, making it a role model for other cities dealing with comparable issues (Jørgensen et al., 2022; Gulsrud et al., 2023). Rotterdam, in the Netherlands, is another noteworthy example of a city that has effectively implemented water management measures for climate resilience. Rotterdam, known around the world for its creative approach to water management, has changed the face of its urban environment to meet the difficulties presented by rising sea levels and more rainfall because of climate change. For efficient water management, the city has combined structural and natural solutions. Using wetlands and natural floodplains to manage excess water, Rotterdam's Room for the River programme integrates carbon neutrality while lowering dependency on energy-intensive pumping equipment. Furthermore, the city's effective water management techniques—like smart water infrastructure and adaptive delta technology—reduce emissions and energy use while boosting climate resilience. Creating multipurpose infrastructure, such as water plazas—public locations that may temporarily hold rainwater during periods of high precipitation to prevent flooding while simultaneously acting as recreation areas during dry spells—is one of the main tactics. To guard against the high tides and storm surges of the North Sea, Rotterdam also has

an advanced system of levees, storm surge barriers, and pumping stations. The concept of "Room for the River," which entails providing greater room for rivers to safely overflow during extreme weather events and lowering the risk of flooding in densely populated regions, has also been embraced by the city. These programmes show how a city can effectively incorporate water management into urban planning to increase catastrophe resistance caused by climate change (van der Berg, 2023; Willems et al., 2023). Another illustration is Singapore, a city-state that, despite its scarce natural water supplies, has emerged as a global pioneer in sustainable water management. Singapore's Four National Taps concept promotes carbon neutrality by using energy-efficient water sources including renewable energy-powered desalination and advanced wastewater treatment via the NEWater system. Furthermore, the city-state's significant rainwater collection and imported water management lessen its need for carbon-intensive water production, which is consistent with its climate and sustainability objectives. The "Four National Taps" concept, which incorporates imported water, desalinated water, local catchment water, and NEWater (recycled water), is the focal point of Singapore's approach to water resiliency. Singapore has made significant investments in desalination and water recycling technologies, which have reduced its reliance on outside water sources and increased its resistance to droughts and other water-related problems. Two-thirds of the city-state's geographical area now serves as a catchment area for collecting and storing rainwater due to the massive rainwater harvesting systems that have been constructed. Furthermore, Singapore has led the way in the application of smart water management technologies, which monitor and optimize water distribution and usage throughout the city. By using these tactics, Singapore has created a resilient and adaptable water delivery system that can deal with present and upcoming climate issues, establishing a standard for other cities hoping to accomplish climate resilience through creative water management (Feldman, 2022; UN.ESCAP, 2022; Adedeji et al., 2022).

ii. Lessons learned from past climate-related disasters

Lessons about the significance of proactive and integrated water management techniques for improving urban resilience have been learned from past climate-related disasters. A crucial case study of the effects of poor water management planning in the face of catastrophic weather events is the Bangkok, Thailand, floods of 2011. Large portions of the city were submerged by the floods, which were among the worst in the nation's history and severely damaged houses, businesses, and infrastructure. Enhancing flood forecasting and early warning systems was one of the most important lessons learned from this event. The impact of the floods was made worse by a lack of timely and accurate information, underscoring the necessity of creating sophisticated monitoring and communication systems to more accurately forecast and manage flood risks. The catastrophe also highlighted the need for coordinated urban design that considers the capacity of current drainage systems as well as natural water flows. Bangkok's experience shows that, especially as climate change increases the frequency and intensity of extreme weather events, neglecting to invest in and maintain vital water infrastructure, such as drainage networks and canals, can have disastrous results (Koh et al., 2022; Loc et al., 2023; Khaspuria et al., 2024). The 2005 hurricane that devastated New Orleans, Louisiana, Hurricane Katrina provides yet another sobering reminder of the difficulties in managing urban water resources and resiliency. The hurricane destroyed New Orleans' levee system, which was not built to resist such a strong storm surge and had not been properly maintained. As a result, the city experienced devastating flooding. The incident brought to light the risks associated with depending on antiquated technologies as well as the vital need to make investments in resilient infrastructure that can survive extreme disasters. Katrina also made clear that improved cooperation between federal, state, and local authorities is required for disaster response and recovery operations. The disaster's human and financial costs were increased by the disjointed reaction and lack of preparation, which also caused major delays in the delivery of help. The necessity of thorough, multi-jurisdictional disaster preparedness and the significance of ongoing investment in resilient infrastructure are two of the most important lessons learned from Hurricane Katrina. Since then, New Orleans has strengthened its flood defences, building the Hurricane and Storm Damage Risk Reduction System, which is intended to withstand a storm 100 years in the future. These advancements show the continued need for awareness and adaptability in the face of climate change and represent the lessons learned from previous mistakes (Kuo et al., 2022; Potutan et al., 2021; Olsonoski & Gianoli, 2024). The 2011 Great East Japan Earthquake and tsunami, which significantly damaged the water infrastructure in the impacted areas, serves as another instructive scenario. The mishap exposed the shortcomings of centralized water systems and emphasized the demand for adaptable, decentralized water management techniques. Following the incident, Japan made investments to modernize its water system, constructing stronger flood defences and encouraging decentralized water management systems. Requirements for water systems to be able to adapt to changing conditions are reflected in the upgrades' emphasis on resilience and flexibility, which highlights the need to be ready for a variety of potential climatic impacts (Goda et al., 2022; Chattha et al., 2022). The value of combining cutting-edge water management techniques and using what has been learned from previous disasters is highlighted by these case studies. Cities can improve their resilience to climate-related calamities and help achieve more general objectives of urban sustainability and carbon neutrality by combining cutting-edge technologies, nature-based solutions, and community-based initiatives, as well as by learning from past events.

5.2.5 Policy and Governance

International cooperation and well-coordinated policy and governance at all governmental levels—from local to national—are essential for efficient water management in urban settings. To ensure that policies are suited to particular requirements while remaining in line with larger national objectives, local, regional, and national governments all play critical roles in the implementation of water management strategies. Building resilient communities that can survive climate-related issues requires the integration of water management strategies into urban planning and disaster preparedness. This integration guarantees that water management is integrated into the larger framework of disaster preparedness and urban development, rather than being in a vacuum. International frameworks and cooperation also offer vital support, facilitating cross-border exchanges of best practices, resources, and information. Cities may develop resilient, flexible water management systems that support urban sustainability and resilience by coordinating efforts at all levels of government and encouraging international cooperation (Lukat et al., 2023; Kalogiannidis et al., 2023; Santos et al., 2023).

i. Role of local, regional, and national governments in water management
   The role of local, regional, and national governments in water management is crucial to ensuring sustainable and resilient urban environments, especially in the context of climate-related disasters. One important example of such an attempt is the Water Framework Directive of the European Union, which lays out a thorough strategy for managing water resources while reducing their negative effects on the environment. To minimize carbon emissions related to water supply and wastewater treatment, this directive encourages member states to implement policies that improve water efficiency, cut pollution, and integrate energy-efficient treatment technology. The regulation guarantees that water management is in line with more general climate and sustainability objectives by establishing legal structures and mandating frequent assessments. (Alevizos et al., 2023). Similarly, the Clean Water State Revolving Fund (CWSRF) in the US offers funding to states, local governments, and municipalities for initiatives that reduce carbon emissions and increase the resilience of water infrastructure. This programme provides money for projects like installing energy-efficient equipment in wastewater treatment plants, incorporating solar or biogas energy into water facilities, and improving stormwater management to control excessive energy usage. To assist local governments and communities in achieving carbon neutrality and enhancing climate resilience, the CWSRF prioritizes projects that lower greenhouse gas emissions and advance sustainable water management (Bade & Tomomewo, 2024). The management of water resources in cities, the supervision of the execution of water infrastructure projects, the upkeep of drainage systems, and the enforcement of local water use laws are frequently the responsibilities of local governments. Through activities like constructing flood defence devices, encouraging water conservation, and interacting with locals to improve their knowledge of water-related concerns, they play a crucial role in tailoring water management methods to the unique requirements and vulnerabilities of their communities. Water management must be incorporated into larger urban planning initiatives by local governments to prevent new construction from increasing the risk of

flooding or putting undue pressure on the water supply. Their role in the rapid response to water-related difficulties is crucial because of their proximity to the population, which enables them to customize water management plans to local conditions (Lukat et al., 2023; Dieperink et al., 2023; Battemarco et al., 2022). Governments from various jurisdictions work together at the regional level to manage water resources that cut across boundaries between states and cities, like aquifers and rivers. To maintain fair distribution of water, safeguard shared water bodies from contamination, and handle water scarcity during droughts, regional governments frequently create and implement regulations. By encouraging cooperation and communication amongst local governments, they also play a critical role in disaster preparedness by guaranteeing that water management plans are uniform and successful throughout the area. On the other hand, national governments offer the overall framework for managing water resources, establishing rules and guidelines that local and regional authorities must abide by. Large-scale water infrastructure projects, such as reservoirs and dams, fall under their purview. They are also in charge of making sure that water management policies support national priorities like climate resiliency and environmental preservation (Daoud et al., 2022; Pokhrel et al., 2022; Anghileri et al., 2024). Along with collaborating with international organizations on transboundary water concerns, national governments are essential in providing funds for and encouraging innovation and research in water management technologies. When these several governmental tiers work together, they provide a multi-layered governance framework that is crucial for efficiently managing water resources, especially in light of climate change.

ii. Integration of water management policies into urban planning and disaster preparedness

Building resilient communities that can withstand and adapt to the difficulties posed by climate change requires integrating water management strategies into urban planning and disaster preparedness. Water management must be considered as a basic element of urban planning, ensuring that choices about land use, infrastructure development, and zoning laws are in line with objectives for disaster risk reduction and sustainable water usage. This integration promotes methods that lessen runoff, improve groundwater recharge, and lessen the urban heat island effect, all of which serve to prevent the escalation of water-related problems like flooding or water scarcity. For instance, integrating green infrastructure—such as urban wetlands, green roofs, and permeable pavements—into city planning not only enhances urban resilience by acting as natural barriers against extreme weather events but also improves water management. It also helps cities achieve carbon neutrality (Lee et al., 2022b; Stoker et al., 2022; Kalogiannidis et al., 2023). On the other hand, to ensure that cities can react swiftly and effectively to water-related emergencies, disaster preparedness necessitates that water management regulations be tightly matched with emergency planning procedures. This includes making sure that water infrastructure is strong enough to endure extreme events and creating backup plans for preserving water supply and quality during emergencies like floods or droughts. Cities may minimize economic disruptions, safeguard public health, and lower the likelihood of catastrophic water failures by including water management in their disaster preparedness plans. Additionally, this integration encourages a more proactive approach to disaster management, in which communities actively seek to lessen their susceptibility to potential hazards while also being ready to respond to situations (Grigg, 2023; Pokhrel et al., 2022; Dąbrowska et al., 2023). To develop comprehensive policies that address both present and future water management concerns, effective integration necessitates collaboration across several sectors and levels of government as well as the involvement of stakeholders, including the corporate sector and civil society.

iii. International frameworks and cooperation
International frameworks and collaboration are imperative to addressing the global challenges of water management in the context of climate-related disasters and urban resilience. Since water resources frequently cross international borders, cooperation between nations is crucial for the management of shared water resources like aquifers, lakes, and rivers. To ensure that water is distributed fairly and that activities in one country do not adversely affect those in another, international frameworks, such as the United Nations Convention on the Law of the Non-Navigational Uses of International Watercourses, provide legal guidelines for nations to manage these shared resources cooperatively. By averting disputes over water resources, which are predicted to increase in frequency as climate change exacerbates water scarcity in many areas, these frameworks support peace and stability (Saikia et al., 2022; Zhao et al., 2022). Sharing best practices, technology, and expertise in water management requires international cooperation as well, particularly when it comes to enhancing resilience to climate-related disasters. Organizations like the World Water Partnership, the United Nations, and the World Bank support water management initiatives in poor nations by supporting them and facilitating the flow of knowledge, enabling them to adopt disaster preparedness and sustainable water practices. Climate resilience initiatives, like the Paris Agreement, encourage nations to embrace water-sensitive measures in their efforts to cut carbon emissions and adapt to the effects of climate change. They highlight the significance of including water management in larger strategies for addressing climate change. International collaboration can also be used to strengthen water governance, provide early warning systems for floods and droughts, and support capacity-building programmes that help cities and regions manage their water resources more sustainably (Sikka et al., 2022; Santos et al., 2023). International frameworks and collaboration are essential in today's increasingly interconnected world to ensure that all countries can manage their water resources effectively in the face of global climate concerns, thereby promoting a more resilient and sustainable future for all.

5.2.6 Innovative Technologies and Approaches

Smart systems, nature-based solutions, and community participation are being combined with innovative technology and approaches in water management to transform urban resilience. Utilizing Internet of Things (IoT) and artificial intelligence (AI)-powered smart water management systems allows for predictive reactions and real-time monitoring, which improves the flexibility and efficiency of water infrastructure. These technologies are a supplement to nature-based solutions that enhance urban landscapes and manage water sustainably, such as green infrastructure and wetlands restoration. Through the involvement of local communities in water management choices and actions, community-based approaches enhance resilience by guaranteeing that measures are culturally appropriate and customized to meet unique needs. The amalgamation of various methodologies results in a complex and robust framework for managing water resources that not only tackles current issues but also fosters sustainable urban development and carbon neutrality throughout time (Orsetti et al., 2022; Nova, 2023; Scheiffer, 2023).

i. Use of smart water management systems (e.g., IoT, AI) for real-time monitoring and response
   An important development in urban water management is the integration of smart water management systems, which make use of technology like AI and IoT to improve resilience and sustainability. Real-time data gathering and monitoring are made possible by these technologies, and they are essential for effectively managing water resources and addressing climate-related issues. To continuously monitor factors like water flow, pressure, quality, and levels, IoT devices, such as smart sensors and metres, can be installed across reservoirs, treatment plants, and water distribution networks. After that, this data is sent to centralized platforms, where AI algorithms examine it to find patterns, foresee any problems, and enhance water management procedures. AI can anticipate demand trends, enhance pumping schedules, and even make proactive system adjustments ahead of extreme weather events, which can minimize the effects of water scarcity or avoid floods (Krishnan et al., 2022; Girotto et al., 2023; Nova, 2023). Urban resilience is further enhanced by smart water management systems, which provide prompt emergency response. IoT sensors can quickly identify problems and notify maintenance staff in the case of a leak or broken pipe, enabling prompt repairs that reduce water loss and avoid damage. AI-driven models may also simulate different climatic scenarios, which aids urban planners in designing infrastructure that can endure future difficulties like heavier rainfall or longer droughts. In addition to increasing operational effectiveness, the use of real-time data and predictive analytics helps long-term planning and sustainability objectives, such as lowering energy usage in water treatment and distribution, which helps achieve carbon neutrality (Bouramdane, 2023; Liu et al., 2022; Ahmad et al., 2022). Smart water management systems will be necessary to provide resilient, adaptive, and sustainable urban settings as cities continue to grow and climate change increases the hazards associated with water.

ii. Nature-based solutions (e.g., wetlands restoration, green infrastructure) in urban water management
Urban water management strategies that rely on natural resources are becoming more widely acknowledged as efficient and environmentally friendly, especially when considering climate resilience and carbon neutrality. In addition to managing water resources and lowering flood risks, these solutions also improve water quality, boost biodiversity, and support the general sustainability of urban areas by using natural processes and ecosystems. For example, restoring wetlands is essential to controlling floods because they function as organic sponges, soaking up extra rain and lowering the amount and speed of runoff that enters cities. Additionally, before contaminants enter rivers and lakes through restored wetlands, they can filter them out, enhancing water quality and promoting the wellbeing of aquatic ecosystems. Wetlands not only help with water management but also store carbon, which helps with attempts to mitigate global warming (Datta et al., 2022). Another important nature-based approach is green infrastructure, which incorporates natural components into urban settings to manage water more sustainably. Green roofs, bioswales, permeable pavements, and urban forests are a few examples of how these features work together to improve groundwater recharge, lessen surface runoff, and lessen the impact of the urban heat island effect. In addition to delivering crucial water management services, these solutions have the added benefits of bettering air quality, increasing green space, and improving urban aesthetics, all of which make cities more habitable and resistant to the effects of climate change. Furthermore, green infrastructure can be more adaptable and affordable than conventional grey infrastructure, including concrete pipes and channels, especially as climate conditions change. Cities can achieve multifunctional landscapes that address sustainability objectives and water management issues simultaneously, in line with carbon neutrality and long-term resilience, by giving priority to nature-based solutions (Kinoshita & Ozaki, 2024; Orsetti et al., 2022; Wang et al., 2023).

iii. Community-based approaches to water management and resilience
Community-based approaches are essential for reaching carbon neutrality because they enable local residents to actively participate in water conservation activities that lessen the need for energy-intensive water treatment and distribution systems. Conventional water management frequently relies on centralized infrastructure, which uses a lot of energy for wastewater treatment, pumping, and purification. However, the need for such energy-intensive procedures declines when communities embrace decentralized water conservation techniques like rainwater collection, greywater recycling, and effective irrigation. This change makes both urban and rural communities more sustainable by reducing carbon emissions related to wastewater treatment and water supply (Vinagre et al., 2023; John et al., 2024). For example, installing rainwater harvesting systems in homes and businesses lessens the need to extract, pump, and treat vast amounts of water from

centralized sources. Greywater recycling also reduces the burden on wastewater treatment facilities, which frequently employ fossil fuel-powered systems, by reusing the water from showers, sinks, and laundry for non-potable purposes like irrigation. Additionally, community-led projects like awareness campaigns and water-efficient landscaping promote behavioural adjustments that help conserve water even more. Together, these initiatives minimize the energy required for water processing by lowering local water use, assisting communities and cities in their transition to carbon neutrality and guaranteeing long-term water security (John & Ajibade, 2024; Kumar & Thakur, 2024). By encouraging effective irrigation techniques, rainwater collection, and greywater recycling, centralized water treatment and distribution systems—which frequently depend on fossil fuels—are directly less necessary. Active community participation in these practices reduces water management energy use and increases local climate change resilience since decentralized systems can better adjust to changing environmental conditions. Additionally, promoting a sustainable culture within communities through education and awareness campaigns can increase the effectiveness of water conservation initiatives and support larger carbon reduction targets (Rodríguez et al., 2021; Santos et al., 2023; Vinagre et al., 2023; Kumar & Thakur, 2024).

The active involvement of local people in the management and protection of water resources is emphasized in community-based approaches to water management and resilience, especially in urban areas that are susceptible to risks associated with climate change. These methods acknowledge that social capital, group effort, and local knowledge are essential to efficient water management, particularly in areas with underdeveloped or overburdened centralized infrastructure and governance. Cities may guarantee that water policies and practices are better suited to the unique requirements and goals of the populace by including communities in water management choices. Examples of this include establishing local water committees or using participatory planning methods. In addition to increasing the efficacy of water management techniques, this inclusive approach instils in community members a sense of accountability and ownership, which promotes more sustainable water use and increased catastrophe resilience connected to climate change (Biswas et al., 2022; Bouramdane, 2023). Furthermore, by enabling locals to take proactive steps in response to water-related issues, community-based water management is essential to fostering resilience at the local level. Examples of community-driven projects that can greatly lessen an area's susceptibility to floods and droughts are rainwater collection, the upkeep of nearby water bodies, and the establishment of shared green spaces. Since these programmes frequently draw on local expertise and resources, they can be flexible and long-lasting even in environments with little funding or resources. Community-based techniques also make it easier to spread strategies for coping with climate change and conserving water, such as restoring natural watercourses or implementing water-efficient technology (Sin-Ampol, 2023; Scheiffer, 2023). Urban water management and resilience frameworks that incorporate these approaches can help cities create more resilient and equitable communities that are better able to handle the effects of climate change. This will also help cities achieve their overall goals of urban sustainability and carbon neutrality.

5.2.7 Challenges and Barriers

The challenges and barriers to efficient water management in urban settings are intricately linked, and the uncertainties surrounding climate change exacerbate the financial, social, and political hurdles. Financial limitations make it difficult to invest in cutting-edge technologies and durable infrastructure, particularly in poor nations. Social considerations, such as fairness and public awareness, make efforts even more difficult because marginalized people frequently suffer higher risks and have less access to dependable water sources. Politically, competing interests and dispersed governance can impede the implementation of critical reforms in water management. The uncertainties surrounding climate change exacerbate these issues, making it more difficult for planners to project future conditions and create workable solutions. Urban areas are increasingly vulnerable to severe water-related risks due to the unpredictability of climate changes. However, the implementation of flexible and adaptive management techniques is often impeded by political and financial hurdles (Obaideen et al., 2022; Santos et al., 2023).

i. Financial, social, and political obstacles to effective water management
   Effective water management in urban areas, especially in the context of climate resilience and sustainability, is often hindered by a range of financial, social, and political obstacles. One of the biggest obstacles is the lack of funding, especially in poor nations and underfunded towns. Modernizing outdated water infrastructure, putting cutting-edge technology into practice, and sustaining current systems might come at an unaffordable cost. Funds for water management projects may be restricted even in affluent areas due to conflicting budgetary priorities, which can result in inadequate system performance, postponed maintenance, and heightened susceptibility to climate-related calamities. Large-scale projects like building new reservoirs, desalination facilities, or flood defences might require a substantial financial outlay, especially when there are no guarantees about the project's return on investment or when the advantages are gradual and not immediately apparent. The financial barrier is exacerbated by the challenge of securing private investment because water management projects generally have low-profit margins (Santos et al., 2023; Natarajan & Bhallamudi, 2023; Bouramdane, 2024). Effective water management is also significantly hampered by social barriers. The effectiveness of sustainable practices and water conservation projects depends on public engagement and understanding, but both are frequently absent. The public's image of water availability frequently differs from the reality of pollution or scarcity, which fuels opposition to conservation measures like higher rates or limitations on water usage. Furthermore, because marginalized populations sometimes lack

access to dependable water supplies or are more susceptible to the effects of climate-related disasters, social injustices can make water management issues worse. Water management practices may so end up benefiting wealthy groups disproportionately while putting poorer ones at greater risk. Political barriers make water management initiatives even more difficult since they frequently result in inconsistent policies and a lack of coordination because water administration is dispersed among several governmental levels. Comprehending long-term sustainability goals over short-term electoral benefits is one example of how political factors can impede the implementation of comprehensive water management methods (Shunglu et al., 2022; Farber, 2023). Achieving the required consensus for action can be challenging when there are conflicts of interest among stakeholders, including those in industry, agriculture, and environmental groups. These conflicts can hinder or derail important water management programmes.

ii. Climate change uncertainties and their implications for planning
The design and execution of water management plans are made more difficult by climate change, which presents a serious obstacle to efforts to promote urban sustainability and resilience. Cities' ability to plan efficiently is hampered by the unpredictable effects of climate change, which include rising sea levels, altered precipitation patterns, and an increase in the frequency and severity of extreme weather events. Conventional water management systems are frequently created using historical climatic data, which might not be an accurate indicator of the future today. Because of this, infrastructure developed to manage water based on historical trends may not be sufficient considering the current environment, increasing the danger of flooding, drought, and other water-related issues. It is difficult for planners to estimate the precise local effects of climate change, which makes it difficult to decide on the right kind and number of interventions, such as the size of flood defences or the capacity of water storage facilities (Vinagre et al., 2023; Marston et al., 2022; Cea & Costabile, 2022). Furthermore, the uncertainty posed by climate change might cause policymakers to become immobilized in their decision-making as they struggle to allocate their limited resources wisely in the face of unpredictable future circumstances. Large infrastructure projects can have increased financial risks because of this uncertainty, which can also make it more challenging to get political and financial backing. Planning is made more difficult by the possibility that climate change may make already-existing issues with water, such as pollution, shortage, and competition for resources, worse. Furthermore, water management systems must be flexible and adaptable to address a variety of potential future scenarios. This calls for novel ideas and cutting-edge technologies, which can be expensive and difficult to put into practice. Water management techniques become even more difficult when planners consider the long-term effects of climate change on water supply and demand, including possible changes in population and land use patterns (Obaideen et al., 2022; Behrmann & Elin, 2024). To address these uncertainties, planning methods must change to become more dynamic and

adaptable so they can adjust to changing circumstances over time. Enhanced cooperation between sectors and levels of governance is also necessary to create resilience in the face of uncertainty.

5.2.8 Future Directions and Recommendations

The effective integration of emerging trends with practical recommendations for politicians, urban planners, and water managers is critical to the future of water management in urban settings. The implementation of integrated water resource management (IWRM) and the emergence of smart technologies signify noteworthy progressions, endorsing comprehensive and data-driven methodologies for water resource management. These tendencies are in line with suggestions to implement water-sensitive urban design (WSUD), create strong legal frameworks, and promote community involvement. Prioritizing adaptable, flexible methods that take these innovations into account will help policymakers and planners make sure that water management practices not only handle present issues but also improve urban resilience to climate change. Cities may accomplish sustainable water management that supports larger objectives of urban sustainability and carbon neutrality by encouraging collaboration across all levels of governance and investing in both nature-based solutions and cutting-edge digital tools (Kamyab et al., 2023; Cui et al., 2021; Bibri et al., 2024).

i. Emerging trends in water management and urban resilience

New developments in urban resilience and water management are indicative of the growing understanding of the necessity of creative and flexible approaches to deal with the multifaceted problems brought on by climate change. IWRM approaches, which emphasize the coordinated development and management of water, land, and related resources to maximize economic and social welfare without compromising the sustainability of essential ecosystems, are becoming increasingly popular. This is one notable trend in the field. IWRM promotes the integration of water management across sectors like agriculture, industry, and urban development by encouraging consideration of the full water cycle, from source to delivery, use, and return. This all-encompassing strategy is especially crucial in urban environments, where conflicting water needs can cause inefficiencies and disputes that are made worse by stressors brought on by climate change (Zeng et al., 2022; Koop et al., 2022).

The growing emphasis on digitization and the application of cutting-edge technologies in water management is another new trend. Real-time monitoring, predictive maintenance, and optimized water distribution are made possible by the increasing prevalence of smart water systems that make use of IoT, AI, and big data analytics. Cities can now more effectively manage their water resources, cut down on losses, and react to crises like leaks or contamination incidents faster, thanks to this technology. Water management is going digital, and this is opening the door to more individualized and demand-responsive services. As a result, customers will be more aware of and involved in their water usage habits, which will lead to increased conservation efforts. Another important trend is the move towards decentralized water management technologies, like localized rainwater collecting and wastewater recycling (Kamyab et al., 2023; Franco et al., 2023). By diversifying their water sources, these systems improve resilience, lessen dependency on large-scale, centralized infrastructure, and promote sustainability by using less energy and resources for the delivery and treatment of water.

ii. Recommendations for policymakers, urban planners, and water managers
Policymakers, urban planners, and water managers must embrace innovative and integrated approaches that tackle the complex issues of water management in the face of climate change if they are to achieve urban sustainability and resilience in the context of carbon neutrality. Integrating the principles of WSUD into all facets of urban planning is one important suggestion. By supporting the use of green infrastructure—such as bioswales, permeable pavements, and green roofs—to manage stormwater sustainably, lower urban heat island effects, and improve biodiversity, WSUD encourages the integration of water management with urban planning. Cities may develop multipurpose spaces that not only efficiently manage water but also enhance the overall resilience and liveability of urban regions by incorporating WSUD into planning processes (Vinagre et al., 2023; Petschek et al., 2024). The creation and execution of strong regulatory frameworks that assist sustainable water management techniques ought to be a top priority for policymakers. This entails establishing stringent guidelines for the economy of water usage, encouraging the repurposing of wastewater that has been treated, and providing incentives for the use of water-saving devices. Sufficient financial instruments, including grants, subsidies, and public-private partnerships, must be combined with effective regulation to promote investment in new technologies and resilient water infrastructure. Moreover, to guarantee that water management plans are egalitarian, logical, and sensitive to local demands, policymakers must promote increased cooperation between various governmental branches, business sector partners, and communities (Santos et al., 2023; Nwokediegwu et al., 2024). Water managers must use more adaptable management techniques. Preparing for a variety of potential future circumstances entails embracing uncertainty and flexibility in water management plans and utilizing methods for risk assessment and scenario planning. To improve their capacity to react quickly to new difficulties, water managers should also make use of digital tools for monitoring and decision-making in real-time. Lastly, encouraging community involvement and engagement in water management procedures is essential (Alaerts et al., 2022; Judd et al., 2023). Water managers can effectively promote policies that are culturally relevant, inclusive, and effective in encouraging sustainable water usage and resilience to climate-related disasters by incorporating local populations in the decision-making process. Together, these recommendations seek to provide metropolitan regions with the means of managing water resources sustainably, promoting resilience, and aiding in the worldwide pursuit of carbon neutrality.

5.3 CONCLUSION

Several crucial ideas come to light when discussing water management in relation to climate-related disasters and urban resilience. These ideas are essential to comprehending and tackling the problems associated with urban sustainability. First, developing communities that are climate resilient requires efficient water management, especially when extreme weather events like storms, floods, and droughts become more common and severe due to climate change. For cities to be prepared for these difficulties, water management must be incorporated into urban design. This can be achieved by using a combination of conventional infrastructure and cutting-edge techniques like smart water systems and natural solutions. Financial, social, and political impediments continue to be major hurdles that must be surmounted with concerted efforts from all governmental levels and with active community involvement. The landscape of water management is changing, as evidenced by emerging trends including the adoption of IWRM and the growing use of digital technology. These patterns highlight the value of a comprehensive strategy that considers the whole water cycle as well as the possibility for cutting-edge technologies to improve responsiveness and efficiency. In addition to stressing the importance of learning from past mistakes and effective implementations, the case studies under examination also show how important it is to build local, regional, and global cooperation to promote resilient water management methods. The summary of these ideas shows how intricately linked and complicated water management is when seen in the larger context of urban sustainability and carbon neutrality.

Building climate resilience and guaranteeing sustainable urban futures require constant adaptation and innovation in water management. Static approaches to water management are no longer adequate as climate change intensifies extreme weather events, modifies precipitation patterns, and creates new difficulties for the supply and quality of water. To respond to these changing and unpredictable situations, cities must constantly modify their water management policies. To achieve this, it is necessary to embrace planning flexibility, make investments in resilient infrastructure, and cultivate an innovative culture that promotes the creation and application of new methods and technologies. Water management helps achieve carbon neutrality by lowering energy use in the distribution, treatment, and delivery of water through the integration of renewable energy sources and energy-efficient technologies. Green infrastructure, decentralized water treatment, and rainwater collection are examples of sustainable practices that improve urban resilience while reducing greenhouse gas emissions. To assist global climate targets and drastically reduce their carbon footprints, cities and industry should prioritize low-carbon water initiatives. The effectiveness and efficiency of water management systems can be greatly increased by innovation. A sustainable and versatile strategy that works with natural processes rather than against them is provided by nature-based solutions, while the integration of smart technology, such as IoT and AI, allows for real-time monitoring and adaptive reactions. Furthermore, water management techniques that are customized to the unique demands and vulnerabilities of urban populations are grounded in local knowledge thanks to community-based approaches. Thus, to create urban resilience in the face of climate change, policymakers, urban planners, and water managers must place a high priority on ongoing learning, adaptation, and cooperation across sectors and scales. They can help the long-term sustainability and carbon neutrality of urban environments by doing this, in addition to shielding cities from the immediate effects of climate-related disasters.

REFERENCES

Abulibdeh, A. (2024). Geospatial assessment of the carbon footprint of water and electricity consumption in residential buildings in Doha, Qatar. *Journal of Cleaner Production*, 445, 141262.

Adedeji, K. B., Ponnle, A. A., Abu-Mahfouz, A. M., & Kurien, A. M. (2022). Towards digitalization of water supply systems for sustainable smart city development—Water 4.0. *Applied Sciences*, 12(18), 9174.

Aghaloo, K., Sharifi, A., Habibzadeh, N., Ali, T., & Chiu, Y. R. (2024). How nature-based solutions can enhance urban resilience to flooding and climate change and provide other co-benefits: A systematic review and taxonomy. *Urban Forestry & Urban Greening*, *95*, 128320.

Ahmad, T., Madonski, R., Zhang, D., Huang, C., & Mujeeb, A. (2022). Data-driven probabilistic machine learning in sustainable smart energy/smart energy systems: Key developments, challenges, and future research opportunities in the context of smart grid paradigm. *Renewable and Sustainable Energy Reviews*, 160, 112128.

Alaerts, G. J., & Kaspersma, J. M. (2022). Facing global transitions in water management: Advances in knowledge and capacity development and towards adaptive approaches. *Water Policy*, 24(5), 685–707.

Alevizos, V., Georgousis, I., & Kapodistria, A. (2023). Toward climate neutrality: A comprehensive overview of sustainable operations management, optimization, and wastewater treatment methods. *Pollutants*, 3(4), 521–543.

Amirzadeh, M., Sobhaninia, S., & Sharifi, A. (2022). Urban resilience: A vague or an evolutionary concept? *Sustainable Cities and Society*, 81, 103853.

Anghileri, D., Pastori, M., Marcos-Garcia, P., Umlauf, G., Crestaz, E., Seliger, R., … & Carmona-Moreno, C. (2024). Global water challenges in sub-saharan Africa and how to strengthen science-policy dialogues on transboundary governance and cooperation. *Journal of Environmental Management*, 365, 121417.

Ayaz, M., Namazi, M. A., ud Din, M. A., Ershath, M. M., & Mansour, A. (2022). Sustainable seawater desalination: Current status, environmental implications and future expectations. *Desalination*, 540, 116022.

Azmeer, A., Tahir, F., & Al-Ghamdi, S. G. (2024). Progress on green infrastructure for urban cooling: Evaluating techniques, design strategies, and benefits. *Urban Climate*, 56, 102077.

Bade, S. O., & Tomomewo, O. S. (2024). A review of governance strategies, policy measures, and regulatory framework for hydrogen energy in the United States. *International Journal of Hydrogen Energy*, 78, 1363–1381.

Battemarco, B. P., Tardin-Coelho, R., Veról, A. P., de Sousa, M. M., da Fontoura, C. V. T., Figueiredo-Cunha, J., … & Miguez, M. G. (2022). Water dynamics and blue-green infrastructure (BGI): Towards risk management and strategic spatial planning guidelines. *Journal of Cleaner Production*, 333, 129993.

Behrmann, O., & Elin, L. (2024). Climate change, water management, and human rights: A comparative study of India and Somalia, and prospects for legal frameworks (Dissertation). https://urn.kb.se/resolve?urn=urn:nbn:se:lnu:diva-127038

Bhardwaj, A., Kumar, M., Alshehri, M., Keshta, I., Abugabah, A., & Sharma, S. K. (2024). Smart water management framework for irrigation in agriculture. *Environmental Technology*, 45(12), 2320–2334.

Bibi & Kara, (2023). Evaluation of climate change, urbanization, and low-impact development practices on urban flooding. *Heliyon*, 9(1), e12955.

Bibri, S. E., Krogstie, J., Kaboli, A., & Alahi, A. (2024). Smarter eco-cities and their leading-edge artificial intelligence of things solutions for environmental sustainability: A comprehensive systematic review. *Environmental Science and Ecotechnology*, 19, 100330.

Biswas, R. R., Sharma, R., & Gyasi-Agyei, Y. (2022). Adaptation to climate change: A study on regional urban water management and planning practice. *Journal of Cleaner Production*, 355, 131643.

Bonato, M., Sambo, B., Sperotto, A., Lambert, J. H., Linkov, I., Critto, A., … & Marcomini, A. (2022). Prioritization of resilience initiatives for climate-related disasters in the Metropolitan City of Venice. *Risk Analysis*, 42(5), 931–952.

Bouramdane, A. A. (2023). Optimal water management strategies: Paving the way for sustainability in smart cities. *Smart Cities*, 6(5), 2849–2882.

Bouramdane, A. A. (2024). Morocco's path to a climate-resilient energy transition: Identifying emission drivers, proposing solutions, and addressing barriers. *Science and Technology for Energy Transition*, 79, 26.

Cea, L., & Costabile, P. (2022). Flood risk in urban areas: Modelling, management and adaptation to climate change. *A Review. Hydrology*, 9(3), 50.

Chattha, M. K., Wei, Z., & Swatuk, L. (2022). Achieving urban water security in Tokyo. In L. Swatuk & C. Cash (Eds.), *The political economy of urban water security under climate change* (pp. 221–248). Springer International Publishing.

Chaurasia, M., Patel, K., Singh, R., & Rao, K. S. (2022). A sustainable approach to combat climate change: Case studies from some urban systems. In R. Bhadouria, S. Upadhyay, S. Tripathi, & P. Singh (Eds.),*Urban Ecology and Global Climate Change* (pp. 297–320). https://doi.org/10.1002/9781119807216.ch15

Chen, J. J., Wang, H. C., Yin, W. X., Wang, Y. Q., Lv, J. Q., & Wang, A. J. (2024). Deciphering carbon emissions in urban sewer networks: Bridging urban sewer networks with city-wide environmental dynamics. *Water Research*, *256*, 121576.

Chen, S., Zhang, L., Liu, B., Yi, H., Su, H., Kharrazi, A., … & Chen, B. (2023). Decoupling wastewater-related greenhouse gas emissions and water stress alleviation across 300 cities in China is challenging yet plausible by 2030. *Nature Water*, 1(6), 534–546.

Council, O. E., Futures, O. W., & Dalgaard, S. (2022). State of Water Justice in Oregon: A Primer on How Oregon Water Infrastructure Challenges Affect Frontline Communities across the State. University of Oregon Scholars' Bank. https://hdl.handle.net/1794/28543

Crawford, S. E., Brinkmann, M., Ouellet, J. D., Lehmkuhl, F., Reicherter, K., Schwarzbauer, J., … & Hollert, H. (2022). Remobilization of pollutants during extreme flood events poses severe risks to human and environmental health. *Journal of Hazardous Materials*, 421, 126691.

Cui, M., Ferreira, F., Fung, T. K., & Matos, J. S. (2021). Tale of two cities: How nature-based solutions help create adaptive and resilient urban water management practices in Singapore and Lisbon. *Sustainability*, 13(18), 10427.

Dąbrowska, J., Orellana, A. E. M., Kilian, W., Moryl, A., Cielecka, N., Michałowska, K., … & Włóka, A. (2023). Between flood and drought: How cities are facing water surplus and scarcity. *Journal of Environmental Management*, 345, 118557.

Daoud, I. Y. H., Dehnavi, S., & Ribbe, L. (2022). Towards good water governance: An analysis of Jordan's National Water Strategy. *Environmental Management*, 69(5), 847–860.

Datta, S., Sinha, D., Chaudhary, V., Kar, S., & Singh, A. (2022). Water pollution of wetlands: A global threat to inland, wetland, and aquatic phytodiversity. In A. K. Rathoure (Ed.), *Handbook of research on monitoring and evaluating the ecological health of wetlands* (pp. 27–50). IGI Global.

Debangshi, U., Ghosh, P., Tiwari, H., Maurya, D. K., & Kumar, M. (2022). Urban resiliency towards climate change. *International Journal of Environment and Climate Change*, 12(11), 2037–2055.

de Oliveira, G. C., Bertone, E., & Stewart, R. A. (2022). Challenges, opportunities, and strategies for undertaking integrated precinct-scale energy–water system planning. *Renewable and Sustainable Energy Reviews*, 161, 112297.

Dieperink, C., Koop, S. H., Witjes, M., Van Leeuwen, K., & Driessen, P. P. (2023). City-to-city learning to enhance urban water management: The contribution of the city blueprint approach. *Cities*, 135, 104216.

du Plessis, A. (2023). Water resources from a global perspective. In *South Africa's water predicament: Freshwater's unceasing decline* (pp. 1–25). Springer International Publishing. https://doi.org/10.1007/978-3-031-24019-5

Farber, D. A. (2023). Inequality and regulation designing rules to address race, poverty, and environmental justice. *American Journal of Law and Equality*, 3, 2–52.

Farooq, O., Sarwar, N., Afzal, S., Mubeen, K., Rehman, A. U., Ahmed, M., & Ahmad, S. (2023). Disaster risk reduction in agriculture. In M. Ahmed & S. Ahmad (Eds.), *Disaster risk reduction in agriculture* (pp. 21–42). Springer Nature Singapore.

Feldman, D. (2022). Water governance: Historical themes and relevance today. In D. L. Feldman (Ed.), *The governance of water innovations* (pp. 62–95). Edward Elgar Publishing.

Franco, P., Kaba, D., Close, S., & Jundi, S. (2023). Introduction of emerging mobility services in rural areas through the use of mobile network data combined with activity-based travel demand modelling. *IET Intelligent Transport Systems*, 17(8), 1509–1524.

Garcia, M., Yu, D., Park, S., Bahambari, P. Y., Iravanloo, B. M., & Sivapalan, M. (2022). Weathering water extremes and cognitive biases in a changing climate. *Water Security*, 15, 100110.

Girotto, C. D., Piadeh, F., Bkhtiari, V., Behzadian, K., Chen, A. S., Campos, L. C., & Zolgharni, M. (2023). A critical review of digital technology innovations for early warning of water-related disease outbreaks associated with climatic hazards. *International Journal of Disaster Risk Reduction*, *100*, 104151.

Goda, K., De Risi, R., De Luca, F., Muhammad, A., Yasuda, T., & Mori, N. (2022). Earthquake-Tsunami risk assessment and critical multi-hazard loss scenarios: A case study in Japan under the Nankai-Tonankai Mega-Thrust. In M. G. Stewart & D. V. Rosowsky (Eds.), *Engineering for extremes: Decision-making in an uncertain world* (pp. 235–254). Springer International Publishing.

Grigg, N. S. (2023). State government roles in water governance: Time for an upgrade. *Public Works Management & Policy*, 28(2), 117–134.

Grzegorzek, M., Wartalska, K., & Kaźmierczak, B. (2023). Review of water treatment methods with a focus on energy consumption. *International Communications in Heat and Mass Transfer*, 143, 106674.

Gulsrud, N., Maurer, M., Olafsson, A. S., Chang, P., Abitz, J., Suhr, N., & Westergaard, S. (2023, July). Relational values as a nature-based solution in Copenhagen, Denmark. In A. Rubbo, J. Du, M. Ramsgaard Thomsen, & M. Tamke (Eds.), *World congress of architects* (pp. 319–331). Springer International Publishing.

Hakovirta, M. (2024). Socioeconomic aspects of climate change in cities and municipalities. In *Carbon neutrality: Follow the money* (pp. 143–156). Springer Nature Switzerland.

Hauashdh, A., Nagapan, S., Jailani, J., & Gamil, Y. (2024). An integrated framework for sustainable and efficient building maintenance operations aligning with climate change, SDGs, and emerging technology. *Results in Engineering*, 21, 101822.

Huang, Y., Lange, E., & Ma, Y. (2022). Living with floods and reconnecting to the water–landscape planning and design for delta plains. *Journal of Environmental Engineering and Landscape Management*, 30(1), 206–219.

Huang, Y., Zhang, J., Ren, Z., Xiang, W., Sifat, I., Zhang, W., … & Li, B. (2023). Next generation decentralized water systems: A water-energy-infrastructure-human nexus (WEIHN) approach. Environmental Science*: Water Research & Technology*, *9*(10), 2446–2471.

Hui, C. X., Dan, G., Alamri, S., & Toghraie, D. (2023). Greening smart cities: An investigation of the integration of urban natural resources and smart city technologies for promoting environmental sustainability. *Sustainable Cities and Society*, 99, 104985.

John, C. K., & Ajibade, F. O. (2024). Exploring the dynamics of WASH services: challenges, enablers, and strategies for improvement. *Discover Civil Engineering*, 1, 79.

John, C. K., & Pu, J. H. (2023). *Household reusable rainwater technology for developing and under-developed countries*. Taylor & Francis.

John, C. K., & Pu, J. H. (2023). Household reusable rainwater technology for developing and under-developed countries. Taylor and Francis Ltd. Routledge. Household Reusable Rainwater Technology for Developing and Under-Devel (routledge.com).

John, C. K., Pu, J. H., Guo, Y., Keating, M., Al-Qadami, E. H. H., Razi, M. A. M., & Hanmaiahgari, P. R. (2024). 3D numerical modelling and laboratory study of flow field induced by a group of submerged vegetations. *Ocean Engineering*, 312, 119038.

John, C. K., Pu, J. H., Moruzzi, R., & Pandey, M. (2021). Health-risk assessment for roof-harvested rainwater via QMRA in Ikorodu area, Lagos, Nigeria. *Journal of Water and Climate Change*. 12(6), 2479–2494, jwc2021025. https://doi.org/10.2166/wcc.2021.025

Jørgensen, G., Fryd, O., Lund, A. A., Andersen, P. S., & Herslund, L. (2022). Nature-based climate adaptation projects, their governance and transitional potential-cases from Copenhagen. *Frontiers in Sustainable Cities*, 4, 906960.

Judd, M., Horne, A. C., & Bond, N. (2023). Perhaps, perhaps, perhaps: Navigating uncertainty in environmental flow management. *Frontiers in Environmental Science*, 11, 1074896.

Junger, L., Hohensinner, S., Schroll, K., Wagner, K., & Seher, W. (2022). Land use in flood-prone areas and its significance for flood risk management—a case study of alpine regions in Austria. *Land*, 11(3), 392.

Kalogiannidis, S., Kalfas, D., Giannarakis, G., & Paschalidou, M. (2023). Integration of water resources management strategies in land use planning towards environmental conservation. *Sustainability*, 15(21), 15242.

Kamyab, H., Khademi, T., Chelliapan, S., SaberiKamarposhti, M., Rezania, S., Yusuf, M., ... & Ahn, Y. (2023). The latest innovative avenues for the utilization of artificial Intelligence and big data analytics in water resource management. *Results in Engineering*, *20*, 101566.

Khalil, L., Abbas, S., Hussain, K., Zaman, K., Iswan, Salamun, H., ... & Anser, M. K. (2022). Sanitation, water, energy use, and traffic volume affect environmental quality: Go-for-green developmental policies. *Plos One*, 17(8), e0271017.

Khaspuria, G., Ranjan, A., Soni, P., & Dadhich, K. (2024). Natural disaster mitigation strategies: A comprehensive review. *Journal of Scientific Research and Reports*, 30(8), 20–34.

Kılkış, Ş., Krajačić, G., Duić, N., & Rosen, M. A. (2023). Sustainable development of energy, water and environment systems in the critical decade for climate action. *Energy Conversion and Management*, 296, 117644.

Kinoshita, T., & Ozaki, T. (2024). Estimation of possible locations for green roofs and bioswales and analysis of the effect of their implementation on stormwater runoff control. *Frontiers in Climate*, 6, 1287386.

Knuijt, M. (2022). The "urban water machine": A landscape-based approach for urban planning. *Journal of Civil Engineering and Architecture*, 16, 109–125.

Koh, R., Babel, M. S., Shinde, V. R., & Mendoza, G. (2022). Towards climate resilient municipal water supply in Bangkok: A collaborative risk informed analysis. *Climate Risk Management*, 35, 100406.

Koop, S. H., Grison, C., Eisenreich, S. J., Hofman, J., & van Leeuwen, K. (2022). Integrated water resources management in cities in the world: Global solutions. *Sustainable Cities and Society*, 86, 104137.

Krishnan, S. R., Nallakaruppan, M. K., Chengoden, R., Koppu, S., Iyapparaja, M., Sadhasivam, J., & Sethuraman, S. (2022). Smart water resource management using artificial intelligence—A review. *Sustainability*, 14(20), 13384.

Krueger, E. H., McPhearson, T., & Levin, S. A. (2022). Integrated assessment of urban water supply security and resilience: Towards a streamlined approach. *Environmental Research Letters*, 17(7), 075006.

Kumar, A., & Thakur, A. (2024). Industrial water conservation by water footprint and sustainable development goals. In R. McLeman & M. Brüntrup (Eds.), *Current directions in water scarcity research* (Vol. 8, pp. 87–117). Elsevier.

Kumar, A., Kapuganti, D. K., & Vohra, R. (2024a). Water demand and quality management. In R. Vohra & A. Kumar (Eds.), *Advanced geospatial practices in natural environment resource management* (pp. 38–66). IGI Global.

Kumar, P., Debele, S. E., Khalili, S., Halios, C. H., Sahani, J., Aghamohammadi, N., ... & Jones, L. (2024b). Urban heat mitigation by green and blue infrastructure: Drivers, effectiveness, and future needs. *The Innovation*, 5(2), 1–22.

Kuo, S. S., Marshall, J. T., & Rowberry, R. (Eds.). (2022). *The Cambridge handbook of disaster law and policy: Risk, recovery, and redevelopment*. Cambridge University Press.

Lakho, F. H., Qureshi, A., Igodt, W., Le, H. Q., Depuydt, V., Rousseau, D. P., & Van Hulle, S. W. (2022). Life cycle assessment of two decentralized water treatment systems combining a constructed wetland and a membrane based drinking water production system. *Resources, Conservation and Recycling*, 178, 106104.

Lam, K. L., Liu, G., Motelica-Wagenaar, A. M., & van der Hoek, J. P. (2022). Toward carbon-neutral water systems: Insights from global cities. *Engineering*, 14, 77–85.

La Vigna, F. (2022). Urban groundwater issues and resource management, and their roles in the resilience of cities. *Hydrogeology Journal*, 30(6), 1657–1683.

Lebu, S., Lee, A., Salzberg, A., & Bauza, V. (2024). Adaptive strategies to enhance water security and resilience in low-and middle-income countries: A critical review. *Science of the Total Environment*, *925*, 171520.

Lee, J., Younos, T., & Parece, T. E. (2022a). Decentralized green water-infrastructure systems: Resilient and sustainable management strategies for building water systems. In T. Younos, J. Lee, & T. E. Parece (Eds.), *Resilient water management strategies in urban settings: Innovations in decentralized water infrastructure systems* (pp. 1–20). Springer International Publishing.

Lee, M., Kim, H., Lee, J. Y., Yang, J. E., & Lim, C. (2022b). A shift towards integrated and adaptive water management in South Korea: Building resilience against climate change. *Water Resources Management*, 36(5), 1611–1625.

Liu, Z., Sun, Y., Xing, C., Liu, J., He, Y., Zhou, Y., & Zhang, G. (2022). Artificial intelligence powered large-scale renewable integrations in multi-energy systems for carbon neutrality transition: Challenges and future perspectives. *Energy and AI*, 10, 100195.

Loc, H. H., Emadzadeh, A., Park, E., Nontikansak, P., & Deo, R. C. (2023). The Great 2011 Thailand flood disaster revisited: Could it have been mitigated by different dam operations based on better weather forecasts? *Environmental Research*, 216, 114493.

Lukat, E., Lenschow, A., Dombrowsky, I., Meergans, F., Schütze, N., Stein, U., & Pahl-Wostl, C. (2023). Governance towards coordination for water resources management: The effect of governance modes. *Environmental Science & Policy*, 141, 50–60.

Manimaran, R. (2025). A comprehensive review of solar assisted technologies in India for clean water and clean energy. *Clean Energy*, *9*(2), 12–36.

Marston, L. T., Abdallah, A. M., Bagstad, K. J., Dickson, K., Glynn, P., Larsen, S. G., ... & Shaffer, K. (2022). Water-use data in the United States: Challenges and future directions. *JAWRA Journal of the American Water Resources Association*, 58(4), 485–495.

Mishra, R. K. (2023). Fresh water availability and its global challenge. *British Journal of Multidisciplinary and Advanced Studies*, 4(3), 1–78.

Mohammadpourlima, N., Lü, X., Nygård, M., & Girgibo, N. (2024, July). The integration of decarbonization and resilience strategies in shaping sustainable cities. In O. Gervasi, B. Murgante, C. Garau, D. Taniar, A. Maria, A. C. Rocha, M. Noelia, & F. Lago (Eds.), *International conference on computational science and its applications* (pp. 291–305). Springer Nature Switzerland.

Mukhopadhyay, A., Duttagupta, S., & Mukherjee, A. (2022). Emerging organic contaminants in global community drinking water sources and supply: A review of occurrence, processes and remediation. *Journal of Environmental Chemical Engineering*, 10(3), 107560.

Muyambo, F., Belle, J., Nyam, Y. S., & Orimoloye, I. R. (2023). Climate-change-induced weather events and implications for urban water resource management in the free state province of South Africa. *Environmental Management*, 71(1), 40–54.

Natarajan, A., & Bhallamudi, S. M. (2023). Implementation challenges for new projects and technologies. *Technological Solutions for Water Sustainability: Challenges and Prospects*, 275–284. https://doi.org/10.2166/9781789063714_0275

Nova, K. (2023). AI-enabled water management systems: An analysis of system components and interdependencies for water conservation. *Eigenpub Review of Science and Technology*, 7(1), 105–124.

Nwokediegwu, Z. Q. S., Obaigbena, A., Majemite, M. T., Daraojimba, O. H., Oliha, J. S., & Dada, M. A. (2024). Review of innovative approaches in water infrastructure: Sustainable desalination and public-private partnerships. *International Journal of Science and Research Archive*, 11(1), 324–332.

Obaideen, K., Shehata, N., Sayed, E. T., Abdelkareem, M. A., Mahmoud, M. S., & Olabi, A. G. (2022). The role of wastewater treatment in achieving sustainable development goals (SDGs) and sustainability guideline. *Energy Nexus*, 7, 100112.

Ojo, B. (2024). Strategies for the optimization of critical infrastructure projects to enhance urban resilience to climate change. *The Journal of Scientific and Engineering Research*, 11, 107–123.

Okeke, E. S., Ezeorba, T. P. C., Okoye, C. O., Chen, Y., Mao, G., Feng, W., & Wu, X. (2022). Environmental and health impact of unrecovered API from pharmaceutical manufacturing wastes: A review of contemporary treatment, recycling and management strategies. *Sustainable Chemistry and Pharmacy*, 30, 100865.

Olsonoski, A., & Gianoli, A. (2024). Water management for flood control in New Orleans: Key factors contributing to institutional inertia. *Cities*, 147, 104793.

Orsetti, E., Tollin, N., Lehmann, M., Valderrama, V. A., & Morató, J. (2022). Building resilient cities: Climate change and health interlinkages in the planning of public spaces. *International Journal of Environmental Research and Public Health*, 19(3), 1355.

Owolabi, T. A., Mohandes, S. R., & Zayed, T. (2022). Investigating the impact of sewer overflow on the environment: A comprehensive literature review paper. *Journal of Environmental Management*, 301, 113810.

Pellegrino, P. R. M., & Ahern, J. (2023). An evolving paradigm of green infrastructure: guided by water. *Planning with Landscape: Green Infrastructure to Build Climate-adapted Cities*, *35*, 51.

Petschek, P., Aung, A. P., Suwanarit, A., & Irvine, K. N. (2024). Integration of building information modeling and stormwater runoff modeling: Enhancing design tools for nature-based solutions in sustainable landscapes. *Sustainability*, 16(9), 3694.

Piemontese, L., Terzi, S., Di Baldassarre, G., Menestrey Schwieger, D. A., Castelli, G., & Bresci, E. (2024). Over-reliance on water infrastructure can hinder climate resilience in pastoral drylands. *Nature Climate Change*, 14(3), 267–274.

Pokhrel, S. R., Chhipi-Shrestha, G., Hewage, K., & Sadiq, R. (2022). Sustainable, resilient, and reliable urban water systems: Making the case for a "one water" approach. *Environmental Reviews*, 30(1), 10–29.

Potutan, G., Arakida, M., & Suzuki, K. (2021). Improving disaster recovery governance: Lessons from the two ad-hoc recovery agencies. *Post-Disaster Governance in Southeast Asia: Response, Recovery, and Resilient Societies*, 63–86.

Raihan, A., Muhtasim, D. A., Farhana, S., Pavel, M. I., Faruk, O., Rahman, M., & Mahmood, A. (2022). Nexus between carbon emissions, economic growth, renewable energy use, urbanization, industrialization, technological innovation, and forest area towards achieving environmental sustainability in Bangladesh. *Energy and Climate Change*, 3, 100080.

Rezvani, S. M., de Almeida, N. M., & Falcão, M. J. (2023). Climate adaptation measures for enhancing urban resilience. *Buildings*, 13(9), 2163.

Richardson, C. M., Davis, K. L., Ruiz-González, C., Guimond, J. A., Michael, H. A., Paldor, A., … & Paytan, A. (2024). The impacts of climate change on coastal groundwater. *Nature Reviews Earth & Environment*, 5(2), 100–119.

Ridha, T., Ross, A. D., & Mostafavi, A. (2022). Climate change impacts on infrastructure: Flood risk perceptions and evaluations of water systems in coastal urban areas. *International Journal of Disaster Risk Reduction*, 73, 102883.

Roba, N. T., Kassa, A. K., Geleta, D. Y., & Hishe, B. K. (2022). Achievements, challenges and opportunities of rainwater harvesting in the Ethiopia context: A review. *Water Supply*, 22(2), 1611–1623.

Rodríguez, C., Sánchez, R., Rebolledo, N., Schneider, N., Serrano, J., & Leiva, E. (2021). Life cycle assessment of greywater treatment systems for water-reuse management in rural areas. *Science of the Total Environment*, 795, 148687.

Rodriguez-Garcia, L., Hosseini, M. M., Mosier, T. M., & Parvania, M. (2022). Resilience analytics for interdependent power and water distribution systems. *IEEE Transactions on Power Systems*, 37(6), 4244–4257.

Saikia, P., Beane, G., Garriga, R. G., Avello, P., Ellis, L., Fisher, S., … & Jiménez, A. (2022). City water resilience framework: A governance based planning tool to enhance urban water resilience. *Sustainable Cities and Society*, 77, 103497.

Santos, E., Carvalho, M., & Martins, S. (2023). Sustainable water management: Understanding the socioeconomic and cultural dimensions. *Sustainability*, 15(17), 13074.

Scheiffer, R. G. (2023). *Nature based solutions for water sensitive communities: Governance challenges and opportunities in the subsystem Vale do Lobo, Campina de Faro aquifer*, Loulé, Portugal (Doctoral dissertation).

Serra-Llobet, A., Kondolf, G. M., Magdaleno, F., & Keenan-Jones, D. (2022). Flood diversions and bypasses: Benefits and challenges. *Wiley Interdisciplinary Reviews: Water*, 9(1), e1562.

Shadbahr, J., Ebadian, M., Gonzales-Calienes, G., Kannangara, M., Ahmadi, L., & Bensebaa, F. (2022). Impact of waste management and conversion technologies on cost and carbon footprint-Case studies in rural and urban cities. *Renewable and Sustainable Energy Reviews*, 168, 112872.

Shunglu, R., Köpke, S., Kanoi, L., Nissanka, T. S., Withanachchi, C. R., Gamage, D. U., … & Withanachchi, S. S. (2022). Barriers in participative water Governance: A critical analysis of community development approaches. *Water*, 14(5), 762.

Sikka, A. K., Alam, M. F., & Mandave, V. (2022). Agricultural water management practices to improve the climate resilience of irrigated agriculture in India. *Irrigation and Drainage*, 71, 7–26.

Sin-Ampol, P. (2023). Community-based adaptation of repeatedly flooded communities in the Yom River Basin (Doctoral dissertation, The Australian National University (Australia)).

Singh, N., Poonia, T., Siwal, S. S., Srivastav, A. L., Sharma, H. K., & Mittal, S. K. (2022). Challenges of water contamination in urban areas. In *Current directions in water scarcity research* (Vol. 6, pp. 173–202). Elsevier.

Singha, B., Eljamal, O., Karmaker, S. C., Maamoun, I., & Sugihara, Y. (2022). Water conservation behavior: Exploring the role of social, psychological, and behavioral determinants. *Journal of Environmental Management*, 317, 115484.

Ssekyanzi, G., Ahmad, M. J., & Choi, K. S. (2024). Sustainable solutions for mitigating water scarcity in developing countries: A comprehensive review of innovative rainwater storage systems. *Water*, 16(17), 2394.

Stoker, P., Albrecht, T., Follingstad, G., & Carlson, E. (2022). Integrating land use planning and water management in US cities: A literature review. *JAWRA Journal of the American Water Resources Association*, 58(3), 321–335.

UN.ESCAP, et al. (2022). A Leveraging science, technology and innovation for low carbon and resilient cities. https://hdl.handle.net/20.500.12870/4358

van der Berg, A. (2023). Climate adaptation planning for resilient and sustainable cities: Perspectives from the city of rotterdam (Netherlands) and the city of Antwerp (Belgium). *European Journal of Risk Regulation*, 14(3), 564–582.

Vinagre, V., Fidélis, T., & Luís, A. (2023). How can we adapt together? Bridging water management and city planning approaches to climate change. *Water*, 15(4), 715.

Vitale, C. (2023). Understanding the shift toward a risk-based approach in flood risk management, a comparative case study of three Italian rivers. *Environmental Science & Policy*, 146, 13–23.

Wan, X., Kuhanestani, P. K., Farmani, R., & Keedwell, E. (2022). Literature review of data analytics for leak detection in water distribution networks: A focus on pressure and flow smart sensors. *Journal of Water Resources Planning and Management*, 148(10), 03122002.

Wang, M., Liu, M., Zhang, D., Qi, J., Fu, W., Zhang, Y., ... & Tan, S. K. (2023). Assessing and optimizing the hydrological performance of Grey-Green infrastructure systems in response to climate change and non-stationary time series. *Water Research*, 232, 119720.

Willems, J. J., van Popering-Verkerk, J., & van Eck, L. (2023). How boundary objects facilitate local climate adaptation networks: The cases of Amsterdam rainproof and water sensitive Rotterdam. *Journal of Environmental Planning and Management*, 66(7), 1513–1532.

Woon, K. S., Phuang, Z. X., Taler, J., Varbanov, P. S., Chong, C. T., Klemeš, J. J., & Lee, C. T. (2023). Recent advances in urban green energy development towards carbon emissions neutrality. *Energy*, 267, 126502.

Wutich, A., Thomson, P., Jepson, W., Stoler, J., Cooperman, A. D., Doss-Gollin, J., ... & Westerhoff, P. (2023). MAD water: Integrating modular, adaptive, and decentralized approaches for water security in the climate change era. *Wiley Interdisciplinary Reviews: Water*, 10(6), e1680.

Zeng, X., Yu, Y., Yang, S., Lv, Y., & Sarker, M. N. I. (2022). Urban resilience for urban sustainability: Concepts, dimensions, and perspectives. *Sustainability*, 14(5), 2481.

Zhang, Q., Smith, K., Zhao, X., Jin, X., Wang, S., Shen, J., & Ren, Z. J. (2021). Greenhouse gas emissions associated with urban water infrastructure: What we have learnt from China's practice. *Wiley Interdisciplinary Reviews: Water*, 8(4), e1529.

Zhao, Y., Xiong, X., Wu, S., & Zhang, K. (2022). Protection of prior and late developers of transboundary water resources in international treaty practices: a review of 416 international water agreements. *International Environmental Agreements: Politics, Law and Economics*, *22*(1), 201–228.

Zozmann, H., Morgan, A., Klassert, C., Klauer, B., & Gawel, E. (2022). Can tanker water services contribute to sustainable access to water? A systematic review of case studies in urban areas. *Sustainability*, 14(17), 11029.

6 Urban Water Management

Strategies, Challenges and Adaptive Solutions

Adeniyi Osaimianmionmwan Arasomwan, Agbugui Marian Onwude, Abhulimen Fran Ehizokhale, Okhamafe Elizabeth Princess, Odozie John Chiwuzulum, and Inobeme Abel

6.1 INTRODUCTION

Water on Earth occurs in various ways, including seas, rivers, water vapor in the atmosphere, ice caps, ice sheets, aquifers, and moisture content of the soil. Together, they constitute resources of the Earth's water. Water serves as the primary pathway by which climate change impacts ecosystems, ultimately affecting the livelihoods and well-being of communities. Among all the natural disasters, water-related events are one of the most devastating. These include Floods (IPCC, 2014; UNISDR, 2015), Droughts (UNCCD, 2017; NASA, 2020), Storms (NOAA, 2020; UNISDR, 2015), Sea-level rise (NOAA, 2020), and water scarcity (UN Water, 2020; World Bank, 2017). Since 1992, floods, droughts, and storms have negatively affected 4.2 billion people (95% of all those affected by disasters) and resulted in $1.3 trillion in damage, which accounts for 63% of total disaster-related losses, according to UN Water and NASA (2020). However, disasters do not recognize borders, and floods frequently have transboundary effects as seen in Figure 6.1.

Due to climate change, extreme precipitation events will very likely become more intense and more frequent by the end of this century (NOAA, 2020). Enhancing adaptability is crucial as climate change impacts water quantity and quality, alters water temperatures, affects water-related ecosystems, and increases the frequency and severity of extreme weather events like floods and droughts.

Effective water management is vital for both humanity and the globe. Effective demand management is crucial for achieving the United Nations' Sustainable Development Goal 6 (SDG 6). As urbanization progresses, tackling environmental challenges is increasingly important. Cutting-edge water management technologies, such as Advanced Metering Infrastructure (AMI), help optimize water use in urban settings (Brears, 2024). More than half of the world's population resides in urban

DOI: 10.1201/9781003518037-9

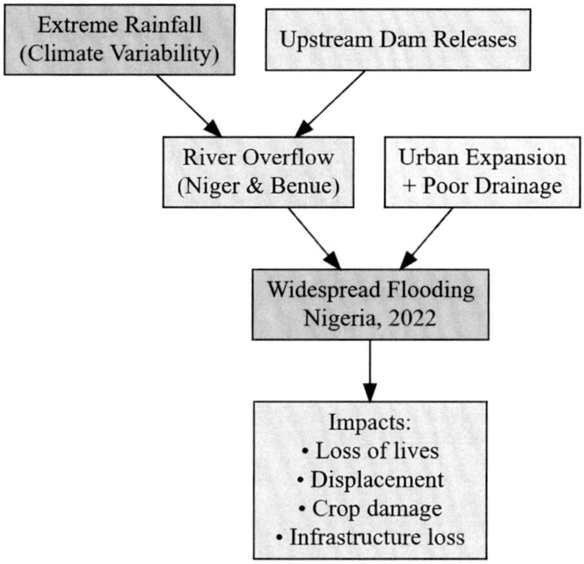

FIGURE 6.1 Flood in Nigeria in the year 2022.

Source: Nigeria Floods affect 1 million—Weather News Aljazeera 2022.

areas, significantly straining water resources. SDG 6 aims to ensure access to clean water and sanitation for all, a goal essential for the sustainable development of cities and the well-being of their inhabitants. Urban waters are essential for cities and play a central role in the shift towards more sustainable green urban environments. Effective management of water resources is vital for enhancing urban climate resilience. As a result, frameworks that combine urban water management with climate change adaptation are becoming increasingly important (Marlow et al., 2013).

A thorough understanding of urban water systems is crucial for preventing significant issues like flooding, water pollution, and shortages (Marlow et al., 2013). Urban water management, alongside the management of water resources in agriculture and industry, is a key focus of the Sustainable Development Goal for Water (SDG 6) which focuses on ensuring the availability and sustainable management of water and sanitation for all. The primary objectives of maintainable city water systems include ensuring hygienic portable water, managing effluents to protect public health, safeguarding against flooding, and mitigating the impacts of contamination (Larsen et al., 2016).

A typical urban water system features an underground network of pipes that supplies drinking water and manages wastewater and stormwater (Figure 6.2). The components at the source include facilities for water collection and storage, as well as aqueducts (such as canals, pipelines, and tunnels) that transport water from its source to treatment facilities. This system encompasses care and supply processes, including storage at the treatment site, as well as collection of effluents (sewer-water) and treatment systems, along with a city system of drainage (Loucks and van Beek, 2017). Drainage infrastructure is crucial for managing floods during heavy rainfall. This

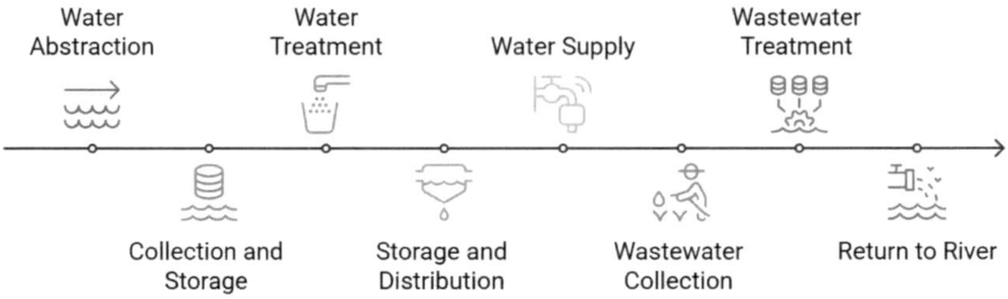

FIGURE 6.2 Diagram of a typical city water system.

Sources: Adapted from Creese and Robinson (1996).

system generates nitrogen-rich wastewater, which can provide nitrogen and other nutrients such as phosphorus, organic matter, and heat, in addition to water.

6.1.1 Water Availability and Resources

About 71% of the surface of the Earth is enclosed by water, which can be classified as groundwater and surface water. Surface water encompasses water from oceans, streams, rivers, ice caps, and glaciers. Freshwater resource comprises precipitation, lakes, rivers, streams, and groundwater. Groundwater, found beneath our feet, is vital for sustaining life, while surface water is primarily used for agriculture and drinking purposes. Groundwater also plays a crucial role in replenishing rivers and lakes. There are over 8,400,000 km³ of freshwater available on Earth, with the majority kept in ice caps and ice sheets in arctic and Antarctic regions. The water cycle plays very vital role in water availability. It facilitates water movement from one location to another, ensuring that it reaches both plants and animals. This cycle is powered by solar energy. However, factors such as deforestation, climate change, and the alteration of waterways can impact the cycle of water and, consequently, the availability of water on the planet.

Resources of water refer to the natural water resources that are beneficial to humans. We primarily utilize water for drinking, food, household tasks, environmental efforts, and industrial activities. Oceans account for 97% of the obtainable water resources. Freshwater comes from underground sources, surface water such as rivers and lakes, and frozen water. Distilled water and treated effluents are considered artificial water resources. Water pollution, climate change, water scarcity, and various human activities significantly impact these resources.

The natural water resource is categorized into surface (superficial) water, groundwater, and frozen water. Surface waters include seas, streams, and rivers. Economically valuable underground water is found in aquifers. Water from ice caps and sheets falls under the category of frozen/ice water resources. Lakes, comprising about 0.008% of freshwater, are created either by the accumulation of snow or by groundwater seeping to the Earth's surface. They can contain both freshwater and saltwater, but freshwater lakes are especially important for human needs. Wetlands consist of 0.0001% of freshwater and are formed in areas where water covers the soil for a long period of time. Deltas, estuaries, marshes, and swamps are wetlands. Wetlands are places where both land-dwelling and aquatic animals can coexist.

6.1.2 Reduction of Water Table

This is the diminution of water which refers to water scarcity, indicating a shortage of water resources. Water scarcity arises when there is insufficient available freshwater to meet demand, impacting every continent. It is characterized by unmet water needs, economic competition over water quality or quantity, conflicts among users, irreversible groundwater depletion, and adverse environmental effects (Figure 6.3).

Half of the world's largest cities face water scarcity. Although only 0.014% of all Earth's water is fresh and readily accessible—while 97% is saline and just under 3% is difficult to access—there is technically enough freshwater on a global scale for humanity to survive.

The key factors responsible for the reduction of Water Table are deforestation and over-pumping of groundwater.

6.1.3 Deforestation

This refers to the process of clearing forestland by cutting down trees and shrubs to transform it into commercial or residential areas. This practice negatively impacts

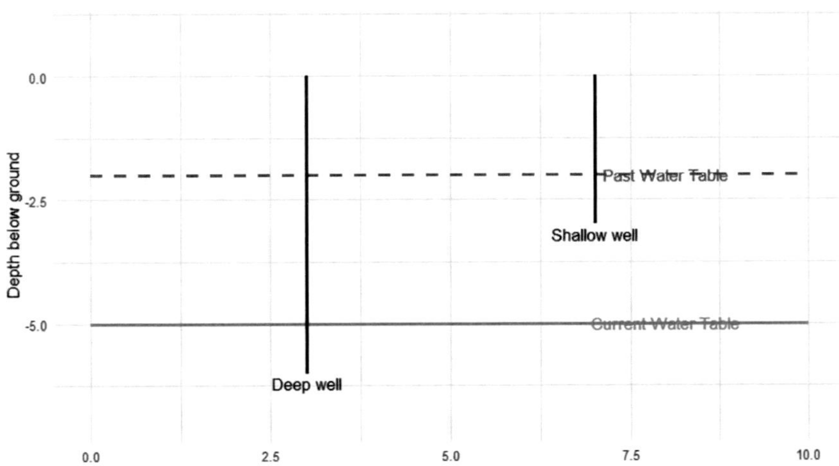

FIGURE 6.3 Conceptual diagram of a depleted water table.

Source: Adapted from Creese and Robinson (1996).

the water table. Woodlands release significant amounts of liquid (water) into the atmosphere through the process of transpiration. As desertification progresses, the area experiences reduced rainfall, with water instead flowing away as river runoff, leading to lasting dryness.

6.1.4 Excessive Groundwater Extraction

Groundwater is water that seeps into the ground and accumulates. It may also be accessed by boring or excavating wells or through pumping. Excessive underground water extraction lowers the water table. In coastal regions, over-pumping can lead to the inland and upward movement of saltwater, resulting in the contamination of the water supply with saltwater.

6.2 URBAN RESILIENCE

Urban resilience includes the following:

i) **Adaptive Capacity**
Adaptive capacity, as defined by UN-Habitat (2018) and the Rockefeller Foundation (2015), refers to the ability of individuals, communities, organizations, or systems to adapt to changing conditions, stresses, or shocks, particularly in the context of urban development and climate change. This emphasizes the importance of adaptive capacity in achieving sustainable urban development, particularly in the face of climate change, rapid urbanization, and other global challenges.

Adaptive capacity encompasses various dimensions:

i. Institutional capacity: The ability of governments, institutions, and policies to support adaptation.
ii. Social capacity: The ability of communities to work together, share knowledge, and build trust.
iii. Economic capacity: The ability to access financial resources, technologies, and innovative solutions.
iv. Physical capacity: The ability to modify or upgrade infrastructure, buildings, and urban planning.
v. Human capacity: The ability of individuals to acquire skills, knowledge, and resilience.

Boosting adaptive capacity allows cities and communities to:

a. Withstand shocks and stresses (such as natural disasters and economic crises).
b. Foresee and prepare for upcoming challenges.
c. Evolve and innovate in response to changing circumstances.
d. Enhance overall resilience and well-being.

ii) **Infrastructure Resilience**
Resilient infrastructure is the capacity of systems and structures to endure and bounce back from negative events, such as natural disasters and climate change. It must not only resist these impacts but also recover swiftly after disruptions. Infrastructure resilience encompasses the ability of physical systems and networks to withstand, adapt to, and recover from adverse occurrences, including natural disasters and other shocks (ASCE, 2020; UNISDR, 2015).

The key principles of infrastructure resilience:

- **Robustness**: The strength and durability of infrastructure to withstand shocks.
- **Redundancy**: Availability of alternative systems or routes to maintain functionality during disruptions.
- **Resourcefulness**: The ability to mobilize necessary resources during crises.
- **Rapid Recovery**: Quick restoration of services and functionality post-disaster.
- **Prevention**: Implementing measures to prevent infrastructure failure before a disaster strikes.
- **Preparedness**: Ensuring systems and communities are prepared for potential impacts through training and planning.
- **Mitigation**: Reducing the potential impact of disasters through better design, planning, and construction practices.
- **Recovery and Reconstruction**: Ensuring that recovery efforts improve resilience and do not simply rebuild what was lost.

The key emphasis here is a proactive approach to building infrastructure resilience, focusing on long-term strategies that integrate risk reduction into infrastructure planning and development. The goal is not just to recover from disasters but to emerge stronger and more capable of withstanding future challenges.

iii) **Community Preparedness**
Community preparedness is essential for reducing vulnerability to disasters and enhancing resilience. Community preparedness involves planning and organizing to ensure that communities can respond effectively to disasters, minimizing risks and enhancing resilience (UNISDR, 2015). This emphasizes the importance of community resilience, advocating for local-level disaster risk reduction strategies that empower communities. Community preparedness is a proactive measure that communities take to anticipate, mitigate, and respond to disasters. Considering the specific needs of vulnerable groups to ensure that preparedness efforts are inclusive and equitable is very important in community preparedness.

Community preparedness involves the following:

- **Risk Assessment**: Understanding local hazards and vulnerabilities to inform preparedness strategies.

- **Community Engagement**: Involving local populations in decision-making processes to **ensure** preparedness measures reflect community needs and knowledge.
- **Education and Training**: Providing training for community members on emergency response, first aid, and evacuation procedures.
- **Resource Mobilization**: Ensuring that communities have **access** to necessary resources, such as **emergency** supplies, communication tools, and training.
- **Collaboration**: Building partnerships with local governments, NGOs, and other stakeholders to create a coordinated approach to disaster preparedness.
- **Awareness and Education**: Raising awareness about disaster risks and providing education on emergency procedures.
- **Community-Based Approaches**: Utilizing local knowledge and capacities to develop tailored preparedness plans.
- **Drills and Simulations**: Conduct regular practice drills to ensure that community members are familiar with response procedures.
- **Emergency Plans**: Develop clear and actionable emergency plans that outline roles, responsibilities, and communication strategies.
- **Psychosocial Support**: Addressing the mental health needs of community members before and after disasters, recognizing the importance of psychological resilience.

The importance of community preparedness is a crucial component of disaster risk reduction. The approaches focus on empowering communities through education, engagement, and collaborative planning, ensuring that local populations are well-equipped to respond to and recover from disasters.

iv) **Economic Resilience:** Economic resilience in urban sustainability refers to a city's ability to withstand and adapt to economic shocks while maintaining sustainable development practices. Integrating economic resilience into urban sustainability efforts not only prepares cities for economic fluctuations but also contributes to long-term environmental and social health. Balancing these dimensions is essential for creating livable, equitable, and thriving urban environments (UN-Habitat, 2018).

v) **Environmental Sustainability:** These are the practices and policies that aim to protect and enhance the natural environment while supporting urban development. Environmental sustainability is a critical pillar of urban sustainability. By integrating ecological considerations into urban planning and development, cities can create healthier, more resilient environments that benefit both current and future generations. Balancing ecological health with economic and social factors is essential for achieving true sustainability in urban areas (Muller, 2007).

6.3 WATER MANAGEMENT STRATEGIES

Water management refers to the regulation and distribution of water resources to mitigate risks to life and property while maximizing their efficient use. Effective management of dams and levees helps decrease the potential for flood damage. Moreover, irrigation water management systems optimize the utilization of limited water supplies in agriculture. Water management encompasses the responsible use and conservation of water resources to promote a sustainable future (Table 6.1). This involves efficient practices in the use, collection, storage, and disposal of water, ensuring reliable supply services, reducing water-related hazards, and protecting the environment.

Water management strategies in urban sustainability focus on efficiently using water resources to support both urban growth and environmental health. Water management involves the strategizing, development, distribution, and efficient use of water resources. As a fundamental necessity, water is essential for all living beings. However, with the increasing scarcity of water, it's crucial to save and manage this vital resource effectively. Here are some key strategies:

1. Flood Prevention/Protection measures: Early warning systems for floods and droughts are very vital for water management (NOAA, 2020; USACE, 2020)
2. Green Infrastructure: Incorporating natural systems, such as green roofs, permeable pavements, and rain gardens, helps manage stormwater, reduce runoff, and improve water quality. These features enhance urban aesthetics and biodiversity as well (US EPA, 2020; The Nature Conservancy, 2020).
3. Water Recycling and Reuse: Implementing systems to treat and reuse wastewater for non-potable purposes (like irrigation or industrial processes) reduces the demand for freshwater sources and minimizes waste (UN Water, 2020).
4. Smart Water Management Systems: Using technology, such as IoT sensors and data analytics, allows cities to monitor water usage in real-time, detect leaks, and optimize water distribution networks, leading to more efficient resource management.

TABLE 6.1
The differences between water resource management and urban water system management

Feature	Water resource management	Urban water system management
Scale	Regional/watershed	Municipal/City
Focus	Natural water bodies and allocation	Urban water infrastructure and services
Main components	Rivers, lakes, aquifers, ecosystems	Water supply, sewers, drainage systems
Objective	Sustainable use and conservation	Safe, reliable water services in cities
Primary stakeholders	Governments, farmers, environmentalists	Municipalities, utilities, urban residents
Examples	River basin planning, aquifer recharge	Water supply networks, wastewater plants

5. Stormwater Management: Developing comprehensive stormwater management plans that include retention basins, bioswales, and detention ponds can reduce flooding and improve water quality by filtering pollutants.
6. Rainwater Harvesting/Management: Encouraging the collection of rainwater from rooftops for irrigation and other uses can supplement municipal water supplies, especially in areas prone to drought (US EPA, 2020b; UN-Water, 2020; World Bank, 2017).
7. Public Education and Engagement: creating awareness about water conservation practices and engaging the community in sustainable water initiatives fosters a culture of responsible water use.
8. Water-Efficient Landscaping: Promoting gardening and the use of native plants in urban landscaping reduces irrigation needs and conserves water.
9. Integrated Water Resources Management (IWRM): Adopting a coordinated approach that considers the entire urban water cycle—supply, distribution, use, and treatment—ensures sustainable management of water resources (FAO, 2023; UN-Water, 2020).
10. Partnerships with Stakeholders: Collaborating with local governments, businesses, and community organizations can enhance resource sharing, innovation, and implementation of sustainable practices.

6.4 ESSENTIAL ELEMENTS OF WATER MANAGEMENT COMPRISE

a. **Policy and Governance**
Addressing cross-border water issues, ensuring fair resource distribution, and promoting sustainable water use can be achieved by creating and enforcing management frameworks, incentives, and regulations at local, national, and international levels.

b. **Water Efficiency and Conservation**
Promoting technologies and practices that minimize water loss and enhance its utilization across various industries. This involves implementing water recycling and reuse systems, retrofitting existing buildings with water-saving fixtures, and creating more efficient irrigation techniques.

c. **Coordinated Water Source Management**
Embracing comprehensive approaches that recognize the interconnectedness of water sources—such as watersheds, lakes, rivers, and aquifers—can help balance competing needs. This ensures that water remains accessible for both present and future generations.

d. **Pollution Regulation and Treatment of Wastewater**
The treatment and management of wastewater safeguards water quality and prevents pollution in aquatic habitats, groundwater, and freshwater sources. This process includes the removal of contaminants/pollutants and the safe disposal or reuse of treated wastewater.

e. **Climate Adaptation and Resilience**
Drought's preparedness, managing floods, and conserving and storing water are essential strategies for building resilience to climate change impacts.

This involves addressing shifting drought patterns, frequent dangerous weather events, intensifying temperatures, and changes in precipitation.

f. **Water Distribution and Supply**
Building infrastructures like water treatment plants, pipelines, and reservoirs are essential for ensuring reliable access to safe drinking water for households, businesses, and agriculture. Effective water management is crucial to tackle issues such as social equity, public health, food insecurity, and biodiversity conservation. By prioritizing fair and efficient water resource management, communities can enhance their resilience to environmental changes and safeguard water supplies for future generations.

6.5 BENEFITS OF MANAGING WATER EFFECTIVELY

Water is essential for maintaining the integrity of the natural environment and serves as a crucial driver of economic and social development.

The accessibility of potable water with its utilization in protection, manufacturing, and food relies on efficiently managing water. It is crucial for economic advancement, environmental viability, and community health.

At its core, managing water involves addressing challenges such as climate change, pollution, water scarcity, and flooding. Achieving equitable and efficient water use requires integrative approaches that integrate community engagement, technology, science, and policy.

Poor water management impacts community parity, human healthiness, and sustainable environmental and economic balance.

The benefits of good water management include:

i. **Economic Stability**
Water plays a critical role in energy generation, particularly for hydroelectric energy and cooling in thermal-energy plants. Managing water effectively ensures reliable energy availability. Additionally, water is vital in various manufacturing processes. Managing water effectively and proper usage is able to improve prices and production (UNISDR, 2015).

ii. **Better Cities**
Municipal zones require efficient water management systems to address stormwater, manage effluent, and supply potable water. The well-being and functionality of these regions rely on such systems. To meet the growing demands of city populations, sustainable management of water strategies is crucial for ensuring both the availability and quality of water. (UN-Habitat, 2018).

iii. **Enhanced Water Quality**
Hygienic water is essential for drinking, cooking, and personal hygiene. Poor management practices can lead to pollution, resulting in waterborne diseases like dysentery and cholera. Effective sanitation is a critical component of water management that promotes public health and reduces the risk of disease transmission (US EPA, 2020a; UN-Water, 2020).

iv. **Reduced Flood Risk**
Implementing effective water management infrastructure can reduce economic losses, safeguard communities, and control flooding. Climate change affects the distribution and availability of water, but proper water management can bolster resilience against these impacts, including changes in precipitation patterns, increased extreme weather events, and rising sea levels (UNISDR, 2015).

v. **Improved Water Supply and Reliability**
Proper management ensures water availability during shortages, aiding in both drought preparation and response. Food is one of the primary uses of freshwater. By implementing efficient irrigation strategies, water management can enhance crop yields and support food security (UN Water, 2020; World Bank, 2017).

vi. **Equity and Accessibility in Society**
Water scarcity can lead to conflicts, but effective management and collaborative agreements can help avoid disputes over water resources. To address inequities, it is essential to manage water resources in a way that ensures equal access for all social groups.

vii. **Supports Urban Biodiversity**
Freshwater resources are limited, and their sustainability for the future relies on good management. Proper management is crucial for maintaining the aquatic habitats, which are essential for biodiversity. Healthy ecosystems support biodiversity by regulating floods, purifying water, and providing habitats for wildlife (UNEP, 2020; The Nature Conservancy, 2020).

6.6 CONCLUSION

Managing water efficiently remains very essential for the effective usage and conservation of water, our world's utmost valuable resource. It encompasses various activities aimed at balancing the demands of different sectors for water while ensuring the ecological integrity of our environments through the management and distribution of water resources. Water management is essential for guaranteeing social justice or fairness, strengthening economies, supporting life, adapting to climate change, and promoting health. Without effective water management, human societies would confront severe environmental and social crises.

REFERENCES

American Society of Civil Engineers (ASCE). (2020). *Together apart: Strength and resilience in the face of adversity*. American Society of Civil Engineers Annual Report.

Brears, R. C. (2024). Water Resources Management: Innovative and Green Solutions *De Gruyter STEM* 2nd Edition, Publisher: Verlag Max Niemeyer, 2024 ISBN: 3111028070, 9783111028071; 180 pages.

Creese, E. E., & Robinson, J. E. (1996). Urban water systems for sustainable development. *Water Resources Journal*, 21(3), 209–220.

Food and Agriculture Organization of the United Nations (FAO). (2023). *Integrated water resources management: A comprehensive approach to water management in agriculture*. Food and Agriculture Organization of the United Nations. https://openknowledge.fao.org/server/api/core/bitstreams/efc7f055-1c0a-485c-9b2a-fef65ee2ef21/content

Intergovernmental Panel on Climate Change. (2014). *Climate change 2014: Synthesis report*. Contribution of Working Groups I, II, and III to the Fifth Assessment Report of the Intergovernmental Panel on Climate Change. IPCC. https://www.ipcc.ch/report/ar5/synthesis-report/

Larsen, T. A., Sabine, H., Christoph, L., Bernhard, T., & Max, M. (2016). Emerging solutions to the water challenges of an urbanizing world. *Science*, 352 (6288), 928–933.

Loucks, D. P. & van Beek, E. (2017). Water resources planning and management. An overview. In D. P. Loucks & E. van Beek (Eds.), *Water resource systems planning and management* (pp. 1–49). Springer. https://doi.org/10.1007/978-3-319-44234-1_1

Marlow, D. R. (2013). Towards sustainable urban water management: A critical reassessment. *Water Research*, 47, 7150–7161. https://doi.org/10.1016/j.watres.2013.07.046

Muller, M. (2007). Adapting to climate change: Water management for urban resilience. *Environment and Urbanization - ENVIRON URBAN*, 19, 99–113. https://doi.org/10.1177/0956247807076726

NASA. (2020). *NASA Earth Observatory: Drought monitoring*. NASA. https://earthobservatory.nasa.gov/features/DroughtMonitoring

National Oceanic and Atmospheric Administration. (2020). *Water management and resources*. NOAA. https:////www.noaa.gov/education/resource-collections/climate/climate-change-impacts

The Rockefeller Foundation (2015)

The Nature Conservancy. (2020). *Nature-based solutions for climate change: Protecting nature to combat the climate crisis*. The Nature Conservancy. https://www.nature.org/en-us/what-we-do/our-insights/perspectives/nature-based-solutions-climate-change/

U.S. Environmental Protection Agency. (2020a). *Climate change and the impact on water resources*. U.S. Environmental Protection Agency. https://www.epa.gov/climate-change

U.S. Environmental Protection Agency. (2020b). *Monitoring and assessment: freshwater – marine*. U.S. Environmental Protection Agency. https://www.epa.gov/water quality

UN WATER (2020). The United Nations World Water Development Report 2020: water and climate change.

UN- Habitat. (2018). *The state of the world's cities* 2018*: The transformation of cities in a generation*. United Nations Human Settlements Programme. https://unhabitat.org/the-state-of-the-worlds-cities-2018

UNISDR (2015). Sendai Framework for Disaster Risk Reduction 2015–2030 (p. 32). United Nations Office for Disaster Risk Reduction (UNISDR). https://www.unisdr.org/we/inform/publications/43291

United Nations Convention to Combat Desertification. (2017). *Drought impact and vulnerability assessment: A framework for improving decision-making in drought-prone areas*. United Nations Convention to Combat Desertification. https://www.unccd.int/publications/drought-impact-and-vulnerability-assessment-framework-improving-decision-making-drought-prone-areas

United States Army Corps of Engineers (USACE) (2020). https://www.nae.usace.army.mil/Missions/Recreation/Mansfield-Hollow-Lake

World Bank (2017). *The City Strength Diagnostic: Promoting Urban Resilience* 17 October 2017 https://www.worldbank.org/en/topic/urbandevelopment/brief/citystrength

7 Emerging Risks of Climate-Related Disasters for Urban Areas

Agbugui Marian Onwude, Okhamafe Elizabeth Princess, Abhulimen Fran Ehizokhale, Inobeme Abel, Odozie John Chiwuzulume, and Adeniyi Osaimianmionmwan Arasomwan

7.1 INTRODUCTION

Global urbanisation has forced people to face the realities of climate change and the natural disasters that accompany it. Cities must immediately increase their resilience against the threats of rising temperatures, unpredictable weather patterns, and rising sea levels, which have ushered in a new era of climate-related calamities. Urban resilience is the capacity of cities to not only withstand challenges but to emerge stronger and more adaptable. However, as urban environments evolve, they also give rise to new hazards that threaten public safety and the integrity of infrastructure. This chapter draws on a variety of academic sources and case studies to examine the complex connections between climate-related disasters, urban resilience, and rising dangers in urban contexts.

7.1.1 Climate-Related Disasters: An Overview

What is disaster? A disaster is a serious disruption to the functioning of a community that exceeds its capacity to cope using its own resources. However, disasters linked to climate change comprise a range of natural occurrences that have the potential to have a substantial effect on both the environment and human systems. These consist of intense heat waves, hurricanes, floods, wildfires, and droughts. There has been a discernible rise in climate-related natural disasters like hurricanes, floods, heat waves, and wildfires. Millions of people have been displaced and their livelihoods have been threatened by these disasters (IPCC, 2022). These catastrophes have the power to uproot communities, seriously impair public health, and severely disrupt economic activities (IPCC, 2021). The frequency and intensity of these disasters have increased significantly over the past few decades, with metropolitan areas being the most affected (UNDRR, 2021; The Guardian, 2024). In China, the soaring

 DOI: 10.1201/9781003518037-10

temperatures, wildfires and drought led to crop failures, power shortages, and factory shutdowns.

Urban regions, with their dense populations and vital infrastructure systems, are most affected by these implications. According to *The Guardian* (2024), in 2024, climate disasters killed 200 people and caused $229 billion in damages. In Nigeria, the devastating floods were likely to be caused by climate crises. *The Guardian* (2024) has reported that the heavy rains behind the recent flooding in Nigeria, Niger and Chad were about 80 times more likely to be crisis related. A clear indication that global heating were impacting the ecosystem and communities. A study by the World Weather Attribution (WWA) group reported that the 2021 drought in the Central Sahel region in Africa damaged crops and contributed to food crisis in 2022 although no estimates of the damage were given.

Urban planning and infrastructure are vulnerable to natural disasters, as seen by the devastating flooding occurrences that have struck cities like New Orleans and Houston in 2021, 2022, and 2024. Miami and other coastal cities are also threatened by increasing sea levels and stronger hurricanes, which makes it more difficult for municipal governments to respond to natural disasters and manage their affairs (Klein et al., 2023).

According to the Intergovernmental Panel on Climate Change (IPCC, 2022), there is no denying the trend that urban areas are becoming more vulnerable due to climate change. Cities face complex problems that jeopardise ecosystems, human health, and infrastructure. The heatwaves that have been documented in locations such as Phoenix, Arizona, and the flooding that has occurred in New Orleans, for example, highlight the serious risks that climatic variability poses (IPCC, 2022).

7.1.2 Causes and Co-Incidences of Climate-Related Disasters

i. **Hurricanes and Tropical Storms**
   Hurricane dangers to coastal towns like Miami and New Orleans are growing as storms become stronger due to warmer ocean waters. One prominent instance is Hurricane Katrina, which struck New Orleans in 2005 and is thought to have killed 1,330 people in addition to causing enormous economic damage. The 2020 hurricane season broke records for the quantity of named storms, and in-depth climate modelling confirms the trend of increasing severity. Hurricanes Katrina in 2005 and Harvey in 2017 are two historical disasters that highlight the severe social and economic effects that these natural disasters had, resulting in protracted recovery times and changes in population dynamics.

ii. **Flooding**
   Heavy downpours, storm surges, or quick snowmelt can cause urban floods, which can affect housing, public transportation, and other vital services. Urban areas are especially vulnerable, according to the IPCC (2022), because of their paved surfaces, which obstruct natural drainage (IPCC, 2022). Unprecedented flash flooding in Washington, D.C. in 2021 brought attention

to how inadequate the current drainage systems are to deal with extreme weather occurrences (NOAA, 2021).

Due to factors including increased urbanisation, climate change, and infrastructure difficulties, urban flooding has become a more urgent problem in many places worldwide. Severe weather events in the past few months have brought attention to how vulnerable metropolitan areas are to floods, necessitating urgent conversations on resilience tactics.

Catastrophic flooding events that have severely damaged houses, infrastructure, and livelihoods have occurred in cities like Mumbai, Houston, and New Orleans in 2021, 2022, and 2024 (The Guardian, 2024). Overdevelopment, insufficient drainage systems, and the impermeable surfaces produced by concrete and asphalt—which hinder water absorption and enhance runoff—all frequently make these situations worse (Pallet et al., 2020).

iii. Wildfires

A bushfire, wildfire, or forest fire is an erratic, unplanned, and uncontrollable fire in a region with flammable vegetation. Another name for a wildfire, depending on the kind of vegetation it is surrounded by, is a bushfire (in Australia), grass fire, hill fire, peat fire, prairie fire, vegetation fire, or veld fire. Wildfires are essential to some natural forest ecosystems. Unlike managed or controlled burning, which is done for human advantage, wildfires are not meant for human use. Severe wildfires sometimes precede rainy climatic cycles that produce large amounts of fuel, which are then followed by heat waves and drought. Climate warming has exacerbated these cycles (Westerling et al., 2006).

In certain places, including Siberia in Russia, California in the United States, British Columbia in Canada, and Australia, wildfires are a frequent calamity. Taiga biomes and regions with Mediterranean temperatures are more vulnerable. People and their communities can be badly impacted by wildfires. Among the effects include the direct health effects of fire and smoke, property damage (particularly in the wildland-urban border), and financial losses. Soil and water contamination is another possibility.

Wildfires are becoming more common in communities next to wildland regions and pose serious dangers to urban infrastructure, public safety, and air quality. The kind of weather that increases the likelihood of wildfires is encouraged by climate change. A rise in wildfires has been directly linked to climate change in certain regions (Parmesan et al., 2022). There is evidence from Earth's past that indicates warmer times had more fire. Evapotranspiration is increased due to climate change. This may lead to soil and vegetation drying out. A fire can spread quickly if it begins in an area with extremely dry vegetation. The fire season may potentially last longer in higher temperatures.

iv. Drought and Water Scarcity

For the most part, drought is a regular occurrence in the climate, with dendrochronological studies dating back to 1900 showing that it is becoming more extreme and unpredictable due to climate change. Environmental, economic, and social effects are the three categories of drought effects. Wetlands drying up, more intense wildfires and a decline in biodiversity are some of the environmental effects.

Urban water shortages have gotten worse, especially in arid and semi-arid areas; this has had detrimental effects on society and the economy. The current drought in California is having a devastating effect on agriculture and water supplies, which means that innovative approaches to water management are required.

One of the most complicated and significant natural disasters, drought has a severe effect on the global economy, ecology, water supplies, agriculture, and society (Jain et al., 2023). Three categories can be used to categorise the effects of droughts and water shortages: environmental, economic, and social (including health).

7.2 THE IMPACT OF CLIMATE-RELATED DISASTERS

The impact of climate change is numerous causing difficulty and unsuitable livelihood to man and his environment. Below are factors resulting from climate change.

i. Health and Social Impacts

Drought has the worst effects on people because it causes crop failure, food crises, starvation, malnutrition, and poverty, all of which increase the risk of death and mass migration (García and Escudero, 1981; Jain et al., 2023).

The health of those who are directly exposed to this phenomenon—excessive heat waves—is negatively impacted. In addition, restricted water supplies, elevated water pollution, elevated food expenses, harvest failure-related stress, lack of water, etc., can all be consequences of droughts. Lower water flows can lead to reduced water quality because they increase the pollution of remaining water sources and decrease the dilution of contaminants (Mosley et al., 2012; Mosley, 2014). This clarifies how factors like water scarcity and droughts contribute to the growing divide between developed and poor nations (Prokurat and Sergiusz, 2015).

Changes in effects are related to vulnerability. For instance, because they lack alternative food sources during a drought, subsistence farmers are more prone to relocate. Populations in areas where water supplies are a primary source of food are more susceptible to starvation.

Low exposure to toxins over an extended period of time can result in cancer due to cytotoxin buildup in food chains and water supplies, some of which are among the most potent toxins known to science. In 2016, freshwater sources throughout the state of California and saltwater seafood from the San Francisco Bay Area were found to contain high amounts of microcystin (San Francisco Bay Environmental News, 2017).

The following are some more instances of social and health effects:

- Water scarcity, crop failure, famine and hunger—drought provides too little water to support food crops; malnutrition, dehydration, and related diseases (García and Escudero, 1981).
- Mass migration, resulting in internal displacement and international refugee.
- Social unrest.
- Conflict over food and water resources.
- Snake migration, which can cause snakebite.

ii. Economic Impacts

Droughts can cause a variety of economic losses, such as decreased productivity in agriculture, forestry, game, and fishing. Drought may also bring about increased costs associated with producing food; decreased energy production in hydro plants; losses resulting from diminished water supplies for tourism and transportation. Other economic losses include issues with water supply for the energy sector and disruptions to municipal water supplies.

Examples of common economic consequences of drought include:

- Habitat degradation, which impacts aquatic and terrestrial wildlife.
- A decrease in the amount of power generated as a result of a decrease in water flow through hydroelectric dams.
- Water shortages for industrial users (Parched village sues to shut tap at Coke/Drought-hit Indians say plant "draining groundwater".

iii. Environmental Impact

Drastic droughts can have a number of negative effects on the environment, such as decreased subsurface and surface water levels, decreased flow levels (with a decrease below the minimum posing a direct danger to amphibian life), increased surface water pollution, drier wetlands, larger and more frequent wildfires, increased deflation intensity, decreased biodiversity, unhealthy trees, and the emergence of pests and dendroid diseases (Adrian et al., 2023).

Droughts have the potential to degrade land and reduce soil moisture, which will destroy farmland productivity (Mekonnen et al., 2022). Reduced yield output, crop growth, and livestock carrying capacity may arise from this. Drought in combination with high amounts of grazing pressure can operate as the tipping point for an ecosystem, generating woody invasion.

Water stress has a number of effects on plant development and quality. To start, a drought can hinder seedling growth and germination (Farooq et al., 2009). Cellular differentiation, expansion, and division are also necessary for plant growth. Due to a decrease in turgor pressure, poor growth is the outcome of impaired mitosis and cell elongation during drought stress. In maize, it has been demonstrated that when water is limited, plant height, biomass, leaf size, and stem girth all drop (Fahad et al., 2017) Drought stress also has a negative impact on crop productivity; a drop in photosynthetic rate, modifications to leaf growth, and a shift in the distribution of resources are all caused by drought stress. Drought-stressed crop plants experience decreases in transpiration rate and leaf water potential. While it decreases for some crops, like potatoes, water-use efficiency rises for others, like wheat (Abbate et al., 2004; Fahad et al., 2017; Kahiluoto et al., 2019).

Drought conditions restrict the ability of plants to absorb nutrients from the soil and distribute those nutrients throughout the plant, which results in stunted development. Drought stress also causes a decrease in photosynthetic activity in plants due to the reduction of photosynthetic tissues, stomata closure, and lower performance of photosynthetic machinery. The decline in plant growth and yields is partially caused by this decrease in photosynthetic activity. The distribution of resources is another aspect that affects decreased plant growth and yields. In response to drought stress, plants will direct more resources towards their roots in order to facilitate water uptake, which will increase root growth and decrease the growth of other plant components while lowering yields (Fahad et al., 2017).

Wind erosion is substantially more severe in arid places and during times of drought. For instance, soil loss from wind erosion in the Great Plains is thought to be up to 6,100 times higher during drought years than during rainy ones (Wiggs and Giles, 2011).

Loess is a fine-grained, silty, fine-coloured, buff or pale yellow, wind-blown (Aeolian), porous, friable, and somewhat cohesive sediment that is generally nonstratified (Richthofen, 1882). Usually, it takes the form of a thick blanket deposit that spans hundreds of square kilometres and occurs widely. Loess frequently forms steep or vertical faces. Soils derived from loess are typically rich in essential minerals such as calcium carbonate, potassium, and phosphates, contributing to their high agricultural fertility. Loess deposits will degrade quickly due to their inherent geological instability, therefore wind erosion can be said to be the major cause of soil loss in loess-derived soils. Thus, farmers frequently plant windbreaks (large trees and bushes) to lessen the wind erosion of loess (Neuendorf et al., 2005; Arthur-Getis et al., 2000; Hofman and Franzen, 1997).

Examples of common environmental consequences of drought include:

- Modification of plant community diversity, which may affect other ecosystem functions and net primary output (Griffin-Nolan et al., 2019).
- Wildfires, like to Australian bushfires and American wildfires, escalate in frequency during periods of drought and have the potential to result in fatalities.
- Dust storms, which occur when a drought strikes an area that is already experiencing desertification and erosion.
- Dust bowls, which are indicators of erosion and exacerbate the landscape's deterioration.
- Snake migration, which can cause snakebite

7.3 EMERGING RISKS FOR URBAN AREAS

Urban environments are subject to new and emerging threats that have the potential to worsen vulnerabilities as they adapt and evolve. Developing comprehensive resilience strategies requires recognising and mitigating these threats.

i. **Increased Frequency and Intensity of Extreme Weather Events**: Hurricanes, floods, heat waves, and wildfires are among the climate-related disasters that are occurring more frequently and with greater intensity in urban areas. These occurrences may result in significant infrastructural damage, community uprooting, and decreased economic output. The Special Report on Climate Change and Cities (2021) published by the IPCC states that cities are especially vulnerable because of their dense population and reliance on linked infrastructures.

   The following are only a few of the negative effects of extreme weather (IPCC, 2021):

   - Too much heat and no rain (heatwave), resulting in droughts and wildfires.
   - Too much rain (heavy downpours), creates floods and landslides.
   - Powerful winds that harm man-made structures and animal habitats, such as hurricanes and tornadoes.
   - Heavy snowfall that can result in blizzards and avalanches.

   The frequency and/or intensity of some extreme weather events have increased due to climate change.

ii. **Urban Heat Islands (UHIs)**: The "urban heat island effect", which is caused by more concrete, asphalt, and less greenery in metropolitan areas, causes temperatures there to often be much higher than in nearby rural areas. The death rate during heatwaves may rise as a result of this effect, especially for vulnerable groups. Urban heat is becoming a bigger health concern, according to the World Health Organization.

   Due to the UHI effect, which intensifies heat waves and raises the energy needed for cooling, cities typically experience higher temperatures than their rural surroundings. The United States Environmental Protection Agency (EPA) (2021) highlights that the UHI effect might result in elevated health risks, especially for susceptible populations like the elderly and low-income inhabitants.

   While extreme heat is frequently accompanied by excessive humidity, it can also be disastrously dry. Heat waves are one of the less well-known types of extreme weather since they are not as obvious as other types of severe weather, such as hurricanes, tornadoes, and thunderstorms (Thornbrugh et al., 2005). Because of the risk of heat cramps, heat expansion, heat stroke, and dehydration or hyperthermia, extremely hot weather can harm crops and communities. Because dried soils are more prone to erosion, there are fewer areas that can be used for farming. Wildfire outbreaks may become more frequent as a result of the increased risk of ignition in dry vegetation.

Plants close their stomata, or leaf holes, in extreme heat as a protective measure to preserve water, but this also reduces the amount of nutrients that plants can absorb. As a result, there is increased air pollution and ozone, which raises population mortality. An estimated 460 people died as a result of increased pollution in the UK during the sweltering summer of 2006 (University of York, 2013) It is predicted that 30,000 more people died as a result of heat stress and air pollution during the summer heat waves in Europe in 2003 (Brücker, 2005). New records for high temperatures have been set in more than 200 American cities (Epstein, 2005). In 1936, the United States had its deadliest heat wave, which claimed the lives of almost 5,000 individuals. In Australia, the deadliest heat wave ever recorded happened in 1938–1939, killing 438 people. 1896 was the second-worst year.

Heat waves can also cause power outages because of the increased demand for electricity (i.e., air conditioning consumption) in those places (Doan and Covarrubias, 2006) The UHI effect can boost temperatures, particularly overnight (Oke, 1982).

iii. **Flood Risk:** Urban areas are more vulnerable to floods as a result of increasing sea levels and more rainfall brought on by climate change. Storm surges and flooding pose a serious threat to coastal cities as they can cause major disruptions to essential services and infrastructure (NOAA, 2021).

Urban areas are especially susceptible to rising sea levels and severe rains since many of them are located in low-lying areas or next to bodies of water. Climate change is anticipated to intensify extreme weather events, resulting in increased frequency and intensity of floods in urban areas, as stated by the IPCC (2022).

Heavy downpours, storm surges, or quick snowmelt can cause urban floods, which can affect housing, public transportation, and other vital services. Urban areas are especially vulnerable, according to the IPCC (2022), because of their paved surfaces, which obstruct natural drainage. Unprecedented flash flooding in Washington, D.C. in 2021 brought attention to how inadequate the current drainage systems are to deal with extreme weather occurrences (NOAA, 2021).

iv. Socioeconomic Disparities: Disasters brought on by climate change disproportionately impact marginalised groups, aggravating already-existing social and economic disparities. Vulnerable groups are frequently the ones most impacted by disasters and take longer to recover, due to their lack of access to resources and support networks. These differences were brought to light by the COVID-19 pandemic, underscoring the critical need for disaster preparedness and equitable urban planning (Wilson et al., 2023). In order to guarantee that all communities are able to recover from the effects of climate change, UNDRR (2020) emphasises the necessity of inclusive planning and resource allocation (UNDRR, 2020).

v. Human Health and Air Quality Deterioration: A study by the Global Climate and Health Alliance found that poor air quality can cause respiratory and cardiovascular issues, which disproportionately affect low-income populations. Climate change can also make cities hotter and more prone to wildfires. Global ecosystems and Earth's climate are intimately linked to air quality. Burning fossil fuels is one of the main causes of greenhouse emissions, which is one of the factors contributing to air pollution (WHO, 2023).

Lung cancer, heart disease, stroke, chronic obstructive pulmonary disease (COPD), respiratory infections, and other pollution-related illnesses are all significantly increased by air pollution. There are 7 million premature deaths annually linked to air pollution (WHO, 2014). An increasing amount of research indicates that exposure to air pollution may be linked to lower IQs, cognitive decline, an increased risk of psychiatric conditions like depression, and poor perinatal health (Allen et al., 2017; Ghosh et al., 2021). Although the impacts of poor air quality on human health are extensive, the respiratory and cardiovascular systems of the body are primarily affected (Lee et al., 2020; Dominski et al., 2021). The type of pollution a person is exposed to, the extent of exposure, the person's health status, and heredity all influence how each person reacts to air pollutants differently (Stanek et al., 2011; Majumder et al., 2022).

vi. **Infrastructure Strain**: Public transit systems, highways, and bridges are examples of urban infrastructure that are frequently not built to endure severe weather. The costs of maintaining and repairing infrastructure will increase dramatically if climate-related disasters become more common (NOAA, 2021).

vii. **Human Health and Vector-Borne Diseases:** Vector-borne illness transmission is associated with climate change. For example, as temperatures rise, the habitats for mosquitoes and ticks grow, increasing the danger of infections such as Zika Virus or Lyme disease. The Centers for Disease Control and Prevention (CDC) has noted that the density and mobility patterns of metropolitan populations may put them at higher risk (CDCP, 2021).

7.4 URBAN RESILIENCE

The "measurable ability of any urban system, with its inhabitants, to maintain continuity through all shocks and stresses, while positively adapting and transforming towards sustainability" is the standard definition of urban resilience (Mariani and Luisana, 2018).

The ability of urban regions to withstand, rebound from, and adjust to shocks connected to climate change is known as urban resilience. Cities that are resilient implement tactics aimed at improving their social structures, infrastructure, and emergency response procedures. Proactive planning is emphasised by the idea, which includes investing in community awareness initiatives, improving public transit, and putting in place green infrastructure (Meerow et al., 2016).

Disasters caused by climate change are becoming more and more dependent on urban resilience. Urban regions' ability to foresee, prepare for, respond to, and recover from crises while maintaining strong community systems is embodied by this. Urban resilience is about using disturbance as a chance to promote long-term sustainability, not merely about "bouncing back" (Rockefeller Foundation, 2016).

To address these issues, cities all across the world are implementing resilience frameworks.

7.4.1 Key Dimensions of Urban Resilience

i. **Physical Infrastructure**
Improving cities' physical infrastructure is essential to reducing the dangers associated with climate change. This can include developing energy-efficient structures, better drainage systems, and flood defences. For instance, Rotterdam has made investments in permeable pavements and green roofs to efficiently manage rainwater and lower urban heat (Zhang et al., 2022). Sydney is pursuing renaturing strategies to combat extreme heat; in Antibes, forests are planted to act as flood barriers; in Medellin, green corridors are enhanced to promote biodiversity; and in Singapore, a dense population of super trees and green corridors are encouraged (Palacios, 2025).

ii. **Ecosystem Services**
As an illustration of how urban landscapes can be changed into safer places, Barcelona's Urban Resilience Strategy is a good example. It aims to minimise flood risks and improve public spaces, combining nature-based solutions with community interaction (BCC, 2023). Nature-based solutions are essential to improving urban resilience and ecosystem health. Green components of a city, such as urban parks and green areas, provide vital ecosystem services including improving biodiversity, filtering air, and reducing UHI effects. Urban greening can yield several advantages at the same time, as demonstrated by the New York City Million Trees initiative (City of New York, 2022).

iii. **Social Systems and Community Engagement**
One of the main components of resilient urban frameworks is community involvement. Residents are better able to endure and react to disasters when they are informed about the hazards involved and appropriate reaction tactics. Engaging local communities in participatory planning procedures strengthens social cohesiveness and increases collective capability. To increase community knowledge and readiness for catastrophes, for example, Los Angeles launched the "Resilience Hubs" initiative (Los Angeles Mayor's Office, 2023). Japan stands out as an international community known to manage disasters effectively, this is because the country experiences many natural disasters due to seismic activities such as typhoons and earthquakes. Thus, this has led to the development of a unique emergency management system with priorities given to disaster prevention and reduction. In essence the 2011 GEJET Machizukuri initiative was introduced (Siawsh et al., 2023).

iv. **Governance and Institutional Capacity**
Resilient urban planning and response capacities are facilitated by efficient governance frameworks. The distribution of resources and the execution of policies depend on multilevel governance, which entails cooperation between local, state, and federal organisations. Strong climate action plans

including numerous stakeholders have been used by cities like Copenhagen to illustrate this (CCAP, 2023). The elements of the CCAP were to make the city the world's best cycle city by 2025 that is CO_2 free thus enhancing livelihood in the city. The city made significant achievements by enabling 1,500,000 people to cycle each day to work thereby increasing the capacity of residents by 60,000.

7.5 CONCLUSION

A multimodal strategy that incorporates creative planning, community involvement, and strong policy frameworks is needed to address climate-related disasters and improve urban resilience. Policymakers and urban planners must be alert and flexible in the face of emerging dangers, especially those pertaining to cybersecurity and socioeconomic inequality.

Resilience-building strategies and sustainable practices must be woven into the urban fabric as we work to address these urgent concerns in order to be ready for a future where climate unpredictability will play an increasingly significant role.

In summary, disasters linked to climate change pose serious hazards to cities, hence quick action to increase resilience is required. By reducing these hazards, effective measures can support sustainable urban growth in the face of escalating climate concerns.

REFERENCES

Abbate, P. E., Dardanelli, J. L., Cantarero, M. G., Maturano, M., Melchiori, R. J., & Suero, E. E. (2004). Climatic and water availability effects on water-use efficiency in wheat. *Crop Science*, 44(2), 474–483. https://doi.org/10.2135/cropsci2004.4740

Adrian, C., Garg, M., Pham, A. V., Phang, S. Y., & Truong, C. (2023). Do natural disasters affect corporate tax avoidance? The case of drought. *Journal of Business Ethics*, *186*(1), 105–135.

Allen, J. L., Klocke, C., Morris-Schaffer, K., Conrad, K., Sobolewski, M., & Cory-Slechta, D. A. (2017). Cognitive effects of air pollution exposures and potential mechanistic underpinnings. *Current Environmental Health Reports*, 4(2), 180–191. https://doi.org/10.1007/s40572-017-0134-3

Barcelona City Council (2023). *Urban resilience strategy: A framework for a sustainable and inclusive city*. Available at: [Barcelona City Council Official Website]

Brücker, G. (2005). Vulnerable populations: Lessons learnt from the summer 2003 heatwaves in Europe. *Eurosurveillance*, 10(7), 1–2. https://doi.org/10.2807/esm.10.07.00551-en

Centers for Disease Control and Prevention (CDC) (2021). *Climate change and human health*. Available at: https://www.cdc.gov/climateandhealth/default.htm

City of New York (2022). *New York City's climate resiliency design guidelines*. Available at: https://www.nyc.gov/site/sirr/index.page

Copenhagen Climate Adaptation Plan (2023). *Strategies for managing water and climate impacts in a collaborative framework*. City of Copenhagen. Available at: https://www.kk.dk/english

Doan, L., & Covarrubias, A. (2006). Heat eases, but thousands of Southern Californians still lack power. *Los Angeles Times*. Available at: https://www.latimes.com/archives/la-xpm-2006-jul-27-me-heat27-story.html

Dominski, F. H., Lorenzetti-Branco, J. H., Buonanno, G., Stabile, L., Gameiro da Silva, M., & Andrade, A. (2021). Effects of air pollution on health: A mapping review of systematic reviews and meta-analyses. *Environmental Research*, 201, 111487. https://doi.org/10.1016/j.envres.2021.111487

Epstein, P. R. (2005). Climate change and human health. *The New England Journal of Medicine*, 353(14), 1433–1436. https://doi.org/10.1056/nejmp058079

Fahad, S., Bajwa, A. A., Nazir, U., Anjum, S. A., Farooq, A., & Zohaib, A. (2017). Crop production under drought and heat stress: Plant responses and management options. *Frontiers in Plant Science*, *8*, 1147. https://doi.org/10.3389/fpls.2017.01147

Farooq, M., Wahid, A., Kobayashi, N., Fujita, D., & Basra, S. M. (2009). Plant drought stress: Effects, mechanisms and management. *Agronomy for Sustainable Development*, 29(1), 185–212. https://doi.org/10.1051/agro:2008021

García, R.V., & Escudero, J. C. (1981). *The constant catastrophe: Malnutrition, famines, and drought*. Pergamon Press. ISBN 9781483189666.

Getis, A., Getis, J., & Fellmann, J. D. (2000). *Introduction to geography*. 7th edn. McGraw-Hill. ISBN 978-0-697-38506-2.

Ghosh, R., Causey, K., Burkart, K., Wozniak, S., Cohen, A., & Brauer, M. (2021). Ambient and household PM2.5 pollution and adverse perinatal outcomes: A meta-regression and analysis of attributable global burden for 204 countries and territories. *PLOS Medicine*, 18(9), e1003718. https://doi.org/10.1371/journal.pmed.1003718

Griffin-Nolan, R. J., Blumenthal, D. M., Collins, S. L., Farkas, T. E., Hoffman, A. M., Mueller, K. E., Ocheltree, T. W., Smith, M. D., Whitney, K. D., & Knapp, A. K. (2019). Shifts in plant functional composition following long-term drought in grasslands. *Journal of Ecology*, 107(5), 2133–2148. https://doi.org/10.1111/1365–2745.13252

Hofman, V., & Franzen, D. (1997). *Emergency tillage to control wind erosion. North Dakota State University Extension Service*. Accessed: 21 March 2009.

Intergovernmental Panel on Climate Change (IPCC) (2021). *Climate change and cities: A special report of the IPCC*. IPCC.

Intergovernmental Panel on Climate Change (IPCC) (2022). *Climate change 2022: Impacts, adaptation, and vulnerability*. IPCC.

Jain, S. K., Wettberg, E. J. V., Punia, S. S., Parihar, A. K., Lamichaney, A., Kumar, J., ... & Toker, C. (2023). Genomic-mediated breeding strategies for global warming in chickpeas (*Cicer arietinum* L.). *Agriculture*, *13*(9), 1721.

Kahiluoto, H., Kaseva, J., Balek, J., Olesen, J. E., Ruiz-Ramos, M., & Gobin, A. (2019). Decline in climate resilience of European wheat. *Proceedings of the National Academy of Sciences of the United States of America*, 116(1), 123–128. https://doi.org/10.1073/pnas.1804387115

Lee, K. K., Bing, R., Kiang, J., Bashir, S., Spath, N., & Stelzle, D. (2020). Adverse health effects associated with household air pollution: A systematic review, meta-analysis, and burden estimation study. *The Lancet Global Health*, 8(11), e1427–e1434. https://doi.org/10.1016/S2214-109X(20)30343-0

Los Angeles Mayor's Office (2023). *Resilience and recovery: A strategic framework for a sustainable Los Angeles*. Available at: https://www.lacity.org/resilience-and-recovery

Machemer, T. (2020). The far-reaching consequences of Siberia's climate-change-driven wildfires. *Smithsonian Magazine*, 9 July. (Accessed: 15 August 2023).

Majumder, N., Kodali, V., Velayutham, M., Goldsmith, T., & Amedro, J. (2022). Aerosol physicochemical determinants of carbon black and ozone inhalation co-exposure induced pulmonary toxicity. *Toxicological Sciences*, 191(1), 61–78. https://doi.org/10.1093/toxsci/kfac113

Mariani, L. (2018). *Urban resilience hub*. Available at: https://urbanresiliencehub.org

Meerow, S., Newell, J. P., & Stults, M. (2016). Defining urban resilience: A review of the literature and its applications to climate change. *Landscape and Urban Planning*, 147, 38–49. https://doi.org/10.1016/j.landurbplan.2015.11.011

Mekonnen, T. W., Mekbib, F., Amsalu, B., Gedil, M., & Labuschagne, M. (2022). Genotype by environment interaction and grain yield stability of drought tolerant cowpea landraces in Ethiopia. *Euphytica, 218*(5), 57.

Mosley, L. M. (2014). Drought impacts on the water quality of freshwater systems: Review and integration. *Earth-Science Reviews*. https://doi.org/10.1016/j.earscirev.2014.11.010

Mosley, L. M., Zammit, B., Leyden, E., Heneker, T. M., Hipsey, M. R., Skinner, D., & Aldridge, K. T. (2012). The impact of extreme low flows on the water quality of the Lower Murray River and Lakes (South Australia). *Water Resources Management*, 26, 3923–3946. https://doi.org/10.1007/s11269-012-0113-2

National Oceanic and Atmospheric Administration (NOAA) (2021). *Coastal flooding*. Available at: https://oceanservice.noaa.gov/hazards/coastal-flooding (Accessed: 5 August 2023).

Neuendorf, K. K. E., Mehl, J. P., & Jackson, J. A. (2005). *Glossary of geology*. Springer-Verlag. ISBN 978-3-540-27951-8.

Oke, T. R. (1982). The energetic basis of the urban heat island. *Quarterly Journal of the Royal Meteorological Society*, 108(455), 1–24. https://doi.org/10.1002/qj.49710845502

Palacios, C. J. (2025). Urban renaturing for a greener future. *Sustainability for All*. Available at: https://www.sustainability.com

Parmesan, C., Morecroft, M. D., & Trisurat, Y. (2022). *Climate change 2022: Impacts, adaptation and vulnerability* (Doctoral dissertation), GIEC. https://hal.science/hal-03774939v1

Prokurat, S. (2015). Drought and water shortages in Asia as a threat and economic problem. *Journal of Modern Science*, 26(3), 235–250.

Rockefeller Foundation (2016). *Urban resilience: Not just about bouncing back*. Available at: https://www.rockefellerfoundation.org/report/urban-resilience-not-just-about-bouncing-back

San Francisco Bay Environmental News (2017). *Toxins from freshwater algae found in San Francisco Bay shellfish*. (Accessed: 5 November 2017).

Siawsh, N., Peszynski, K., Vo-Tran, H., & Young, L. (2023). Toward the creation of disaster-resilient communities: The Machizukuri initiative – The 2011 Tōhoku Great East Japan Earthquake and Tsunami. *International Journal of Disaster Risk Reduction*, 93, 103961. https://doi.org/10.1016/j.ijdrr.2023.103961

Stanek, L. W., Brown, J. S., Stanek, J., Gift, J., & Costa, D. L. (2011). Air pollution toxicology—A brief review of the role of the science in shaping the current understanding of air pollution health risks. *Toxicological Sciences*, 120(Suppl 1), S8–S27. https://doi.org/10.1093/toxsci/kfq367

The Guardian (2024). *Climate-related disasters*. Available at: https://www.theguardian.com

Thornbrugh, C., Hzia-Redsteer, M., & Stevermer, A. (2005). The assessment of sand dune mobility from 1980 through 2004 on the Moenkopi Plateau of the Navajo Nation. *NCAR Report*, 99439.

U.S. Environmental Protection Agency (EPA) (2021). *Urban heat island basics*. Available at: https://www.epa.gov/heatislands

United Nations Office for Disaster Risk Reduction (UNDRR) (2020). *Disaster risk reduction in the context of climate change*. Available at: https://www.undrr.org

United Nations Office for Disaster Risk Reduction (UNDRR) (2021). *Global assessment report on disaster risk reduction 2021: For a resilient future*. Available at: https://www.undrr.org/publication/global-assessment-report-disaster-risk-reduction-2021

von Richthofen, F. (1882). On the mode of origin of the loess. *Geological Magazine* (Decade II), 9(7), 293–305. https://doi.org/10.1017/S001675680017164X

Westerling, A. L., Hidalgo, H. G., Cayan, D. R., & Swetnam, T. W. (2006). Warming and earlier spring increase western U.S. forest wildfire activity. *Science*, 313(5789), 940–943. https://doi.org/10.1126/science.1128834

Wiggs, G. F. S. (2011). Geomorphological hazards in drylands. In D. S. G. Thomas (Ed.), *Arid zone geomorphology: Process, form and change in drylands*. 3rd edn. (p. 588). John Wiley & Sons, ISBN 978-0-470-71076-0.

Wilson, T. (2023). The social dimension of climate resilience: Lessons from the COVID-19 pandemic. *Urban Studies Journal*. https://clok.uclan.ac.uk/id/eprint/47077/

World Health Organization (2014). *7 million premature deaths annually linked to air pollution*. Available at: https://www.who.int

World Health Organization (2023). *Air pollution*. Available at: https://www.who.int

Zhang, Y., Wang, J., Xu, L., & Qiu, Z. (2022). Physical infrastructure under key dimensions of urban resilience. *Sustainable Cities and Society*, 78, 103609. https://doi.org/10.1016/j.scs.2021.103609

8 Transformative Adaptation for Climate Change

Assessment of Farm Households' Capacities for Adaptive and Transformative Actions under Climate Change in Nigeria

Agbugui Marian Onwude, Odozie John Chiwuzulum, Inobeme Abel, Abhulimen Fran Ehizokhale, Okhamafe Elizabeth Princess, Adeniyi Osaimianmionmwan Arasomwan, and Ogbiti Temitope John

8.1 INTRODUCTION

The debate on transformative responses by human systems has advanced in many reports of the IPCC particularly in the Fifth Assessment Report (IPCC, 2012, 2014). While there are many dimensions of the debate, fundamentally it includes:

(1) Capacity: the conceptualization of transformative responses as it relates to agriculture (Adger and Bernett, 2009; Howden et al., 2007; IAASTD, 2009; Kates et al., 2012).
(2) The underlying pathways or processes (Fazey et al., 2015; Mapfumo et al., 2014).
(3) The drivers, indicators, and measurement of transformational capacity (Dowd et al., 2014; Marshall et al., 2014; Warner et al., 2015).

Understanding the underlying process in reducing systems exposure and sensitivity to climate risk matters for short- and long-term vulnerability programming.

DOI: 10.1201/9781003518037-11

This chapter focuses on the socio-economic pathways underlying transformative and adaptive actions in Nigeria's agriculture.

Previous studies have mainly focused on incremental adaptation actions viewed as short term and reactive. Yet there is a scant analysis of the social pathways underlying actions of adaptation against vulnerability to climate change. Agriculture plays an important role in Nigeria's economic growth and accounts for employment, foreign exchange earnings, and domestic food production. Agriculture is expected to be impacted by climate change and given the problems of uncertainty, the amplification of climate variability, drought, and flooding, understanding transformative responses becomes critical. However, identifying transformational actions from adaptive or incremental actions has been practically unclear (Copley et al., 2015). A distinction has been drawn between capacity to adapt and actions of adaptation (Campbell et al., 2014). Adaptation measures vary in the way they can reduce vulnerabilities ranging from an incremental or moderate adjustment in existing practices and behaviours to a more fundamental adjustment referred to as 'transformational adaptation' or simply 'transformation' (Asprone and Manfredi, 2014; Engle, 2017; Garschagen, 2015). The role of adaptive capacity as a driver of adaptive and transformative actions has been recognized in a couple of studies (Smit and Wandel, 2006; Marshall et al., 2012; Rickards and Howden, 2012). Previous studies analysed adaptive capacity using cross-country data (Kelly and Adger, 2000; Yohe and Tol, 2002; Brooks and Adger, 2004; Adger et al., 2004; Haddad, 2005; Brooks et al., 2005; Vincent, 2007). There are only a few studies that applied a theoretical framework (Williges, 2016). The remainder of this chapter is divided into five sections. Section 8.2 gives a review of the literature while Section 8.3 is the methodology. Section 8.4 presents the analytical framework, Section 8.5 presents the results, and Section 8.6 presents the conclusion.

8.2 LITERATURE REVIEW

Adaptive and transformative responses depend on the adaptive capacity of the system (Park et al., 2012). For example, transformative responses can be livelihood diversification or active engagement in long-term planning while adaptive responses can be a range of behavioural changes. While a substantial analysis of human responses to climate change is focused on actions whether adaptive or transformative (Barnes et al., 2020), there is scant information as to the weight of adaptive capacity needed for transformative and adaptive actions. The complexity of adaptive capacity makes it an indeterminate concept. It is defined as "the ability of a system to adjust, modify, or change its characteristics or actions to moderate potential damage, take advantage of opportunities, or cope with the consequences of shock or stress" (Jones et al., 2010). As noted in Smit and Wandel (2006), the following characteristics are associated with adaptive capacity. First, it is dynamic and determined by factors at the local, national, and global scales. The factors are also interdependent. Secondly, the system's adaptive capacity is context-specific and path-dependent. Thirdly, it has a coping range and limits; that is, while adaptation is possible within moderate deviations from the norm, for extreme events adaptive capacity may lie outside the coping range or may exceed the adaptive capacity of the community. Fourthly, the system's adaptive capacity and coping range are not static but respond to changes in economic, social, political, and institutional conditions over time.

For an empirical analysis, there is a challenge of how to identify the indicators of adaptive capacity underlying adaptation and transformative actions. However, several frameworks have been developed to operationalize the concept of adaptive capacity, such as the entitlement, socio-economic, and sustainable livelihood frameworks (SLFs). The Entitlement framework draws from socio-economic and institutional analysis of vulnerability and predicts adaptive capacity to be determined by governance, politics, institutions, and culture. The socio-ecological framework draws from ecological economics and considers the activities of humans as embedded within complex socio-ecological systems (Gallopín, 2006; Nelson et al., 2010). The SLF is a combination of entitlements and rural livelihood frameworks and predicts adaptive capacity to depend on people's access to assets such as natural, human, social, physical, and financial capital and the mediating factors of institutions. Other frameworks use assets, institutions, entitlements knowledge, information, innovation, and flexible forward-looking decision-making as dimensions of adaptive capacity. Similarly, some studies identify assets, flexibility, organization, learning, socio-cognitive constructs, and agency as determinants of adaptive capacity (Cinner and Barnes, 2019; Cinner et al, 2018; Barnes et al., 2020).

8.3 METHODOLOGY

The procedure developed for measuring the adaptive capacity of farmers and aggregated across zones in Nigeria is presented here and consists of two phases. A methodological sequence has been provided in Table 8.1.

The first requires identifying an appropriate framework or theory to guide the selection of indicators. Frameworks help to establish the boundary of the proximate factors driving the adaptive and transformative capacities of a system under study. Several studies have used the SLF to identify indicators for adaptive capacity analysis (Reddy et al., 2004; Babulo et al., 2008; Aulong et al., 2011). SLF considers the multidimensionality of adaptive capacity by helping to identify a variety of social and economic indicators. These indicators were sourced from the General Household Survey (GHSPanel) data set collected by the National Bureau of Statistics between 2010 and 2012. The sample size is 5,000 households. Data cleaning was carried out and further reducing the sample size to 1,778 households. The selection of indicators was based on the policy relevance of indicators and to the extent that the data set used allowed. The selection included both quantitative and qualitative indicators measured at different scales (individual, household, and community). Indicators of human capital included the proportion of literate households, years of education, number of persons per household, and the age of the headed household. These measures were included to gauge the ease with which households can access information, health practitioners, and government extension services, their knowledge and skill in enterprise activities. Distance to health facilities is included as a health variable. Indicators of natural capital consist of the proportion of households with tree crops, livestock, and access to water sources both underground and surface. Several indicators of physical capital

TABLE 8.1
A methodological description for a typical transformative adaptation for climate change considering assessment of farm households' capacities for adaptive and transformative actions under climate change in Nigeria

Phase 1	Step 1	
Screening and classification	Identifying an appropriate framework	Primary classification using selective guide to indicators
	Step 2	The use of Sustainable Livelihood Framework (SLF)
	Criticality and vulnerability of variables	
	Step 3	Sample size selection of more than 5,000
	Narrative	
	Step 4	Qualitative data assessment
	Qualitative analysis	
	Step 5	Quantitative data assessment
	Quantitative analysis	
Phase 2	Step 1	Cleaning and analysis of data
Operationalization of framework and indexing off selected indicators using Principal Component Analysis (PCA)	The multiple regression approach uses all the indicators as unconstrained variables to measure weight	
	Step 2	Quantitative data assessment
	Comparisons using PCA variables	

were included, namely the proportion of households with machinery, mobile phones, access to seeds, fertilizer, draught animal traction, farm equipment, acreage planted, and access to electricity.

Access to credit, off-farm income, income from farm produce sales, remittances, and safety nets from the government were used as indicators of financial capital. Different farm households will have different access to official and informal rural credit, which will make it harder for them to get the money they need to buy fertilizer, seeds, and other inputs. Off-farm income, like money sent back to family and friends and wages, is a big part of people's lives in rural Nigeria. It is normal for people who live in rural areas to get money from family and friends. Households also get support from government bodies, NGOs, and formal risk-bearing bodies like the National Agricultural Insurance Agency, the National Food Reserve Agency, and the National Meteorological Services. Social capital includes things like the number of groups in a community, the number of members in each group, the number of women, and the number of people under 30 years old. The number of women members shows how women are treated in decision-making and how many women and young people join self-help groups. These kinds of measures show fairness in terms of benefits and coverage.

8.3.1 Indicator Scrutiny, Selection, and Indexing Using Principal Component Analysis

The second procedure of the methodology is presented here and entails the operationalization of the framework and indexing of selected indicators. The principle underlying the scrutiny and indexing of adaptive capacity indicators is akin to that of multidimensional welfare indices (see Decancq and Lugo, 2013). Given m indicators that are relevant to assessing the adaptive capacity of a farmer or a community. Also given that the adaptive capacity in the selected indicators can be measured in an inter-personal comparable way. Let x_j^i denote the non-negative adaptive capacity of a farmer or community i in indicator j and let the adaptive capacity vector $x^i = \left(x_1^i,\ldots,x_m^i\right)$ summarise these capacities across all m indicators for farmer i. The distribution matrix $X = \left(x^1,\ldots,x^n\right)'$, collects the capacity vectors of the n farmers or communities in a given population. The indexing problem becomes that of searching for an appropriate index that maps the capacity across the vector of indicators so that they can be ranked naturally. The problem is expressed as:

$$I(x) = \begin{cases} \left[w_1 I_1\left(x_1\right)^{\beta} + \cdots + w_m I_m\left(x_m\right)^{\beta}\right]^{1/\beta} & \text{for } \beta \neq 0 \\ I_1\left(x_i\right)^{w_i} * \ldots * I_m\left(x_m\right)^{w_m} & \text{for } \beta = 0 \end{cases} \tag{1}$$

Where index $I(x)$ is the weighted mean of order β of the transformed indicator $I_j\left(x_j\right)$. The parameter β is related to the degree of the elasticity of substitution between indicators. In practice, the standard weighted arithmetic mean of the following form is applied:

$$I(x) = w_1 I_1\left(x_1\right) + \cdots + w_m I_m\left(x_m\right) \tag{2}$$

Equation (2) says that index $I(x)$ is the weighted mean of order $\beta = 1$ of the transformed capacity. $I_j\left(x_j\right)$ There are many approaches for weighting indicators. Decancq and Lugo (2013) and Filmer and Pritchett (1998) identified the subjective approach of assigning weights based on personal, group, or expert judgement. The hybrid approach combines the statistical and subjective approaches. The multiple regressions approach uses all the indicators as unconstrained variables to generate weights. A common statistical approach for weighting is the Principal Component Analysis (PCA). This article employed PCA to weight and aggregate indicators into a composite value.

PCA considers the pattern of correlation between indicators. This is the merit PCA has over the subjective approach of assigning weights. PCA also has the advantage of being able to reduce the many indicators into a few indicators. Most of these indicators are dynamic and interdependent thus reducing the concern for double counting. Finally, standardize indicators to lie between 0 and 1. The drawbacks include, first, the difficulty in interpreting the results. Secondly, PCA can lead to counter-intuitive weights by assigning negative weights to valuable indicators (World Economic Forum, 2001). Bearing these challenges, PCA requires several trials.

PCA was executed in Stata 11. The analysis generated several principal components (PCs) and the eigenvalue parameter. To identify the most important PC to retain, the Kaiser Criterion conventional practice of eigenvalue >1 was used. PCs with variance equal to or greater than 1 were retained for analysis (Filmer and Pritchett, 2001). Thus, in the first run, the eigenvalue and communality of variables guided the selection of indicators. The communality measures the proportion of variance explained in a particular indicator included. Indicators with a high proportion of variance explained were selected. This reduced the number of selected indicators to 15 out of the 42 indicators for the second PCA run. The analysis also generated a matrix of factor weights. This was used to evaluate the relative importance of each indicator selected and shed information on the best indicators defining retained PCs.

8.4 RESULTS AND DISCUSSION

Table 8.2 presents the summary statistics of the 15 indicators included in the second run while Table 8.3 presents the attributes of the five PCs retained for further analysis. These five retained components have eigenvalues greater than 1 and together account for 57% of the variability in the original data. Table 8.4 presents the structure of the weights of the indicators across the retained five PCs. The PCs can be seen as factors predicting the weight of each of the indicators. The weighting process assigns each indicator to the PC that contributes most to explaining the variance of that indicator. For this analysis weights smaller than 0.3 were omitted to facilitate a clean presentation.

From Table 8.4, PC1 predicts household size (0.47), animal traction (0.44), income from crop sales (0.38), and access to forest trees (–0.53) as relatively important indicators across farm households. PC1 most probably predicts an adaptive capacity index that is built on family labour and natural resources. This type of adaptive capacity index was highest for states in Northeast and Northwest ecological zones with an index ranging from 0.52 to 0.6. This is depicted in the map (Figure 8.1).

PC2 predicts membership of an association (0.58), ownership of livestock (0.53), access to commercial fertilizer (0.42), and access to forest trees (0.26). PC2 most probably predicts an adaptive capacity index that relies on social and financial capital. This was highest for states in Northwest ecological zones with an index ranging from 0.52 to 0.7. PC3 predicts access to farm equipment (0.65), available land for cultivation (0.48), access to herbicide (0.47), and crop income (0.28). PC3 most probably predicts adaptive capacity that relies on physical assets most especially farm equipment and land. This is most important in the southeast and southwest ecological zones. PC4 predicts literacy rate (0.63), age (0.50), and subsidized fertilizer (0.44). PC4 most probably predicts the importance of human capital in terms of education and experience. The adaptive capacity index is also built on the extent to which farm households have access to subsidized fertilizer. This is most pronounced in the whole of Northern Nigeria comprising Northeast, Northwest, and North central. PC5 predicted agricultural insurance (0.67) and wage income (0.64). PC5 can be viewed as

TABLE 8.2
Summary statistics of selected indicators

Selected indicators	N	Mean	SD	Min	Max
Area planted	1765	2.066072	2.104359	0	8.695388
Animal traction	1708	.2505855	.4334771	0	1
Farm equipment	1734	.1885813	.3912885	0	1
Subsidized fertilizer	1765	.0226629	.1488685	0	1
Commercial fertilizer	1765	.3524079	.4778556	0	1
Herbicide	1765	.1303116	.3367414	0	1
Livestock ownership	1765	.6458924	.4783775	0	1
Ownership of Economic trees	1765	.1631728	.369628	0	1
Income from crop sales	1765	.5450425	.5003792	0	2
Insurance	1764	.0090703	.0948321	0	1
Wage income	1765	.0209632	.1433016	0	1
Membership of an association	1765	.8430595	.3638478	0	1
Household size	1765	5.815975	1.570156	2	11.6
Age	1760	51.26534	15.30325	15	108
Literacy	1763	.5530346	.4973204	0	1

Source: Author's classification

TABLE 8.3
Principal components and the associated eigenvalue

Principal components	Eigenvalue	Differences	Proportion	Cumulative
1	2.44722	1.06652	0.1771	0.1771
2	1.3807	.0493904	0.106	0.2831
3	1.33131	.0764962	0.1028	0.3859
4	1.25481	.118586	0.0977	0.4836
5	1.13622	.128505	0.0897	0.5733

Source: Author's classification

TABLE 8.4
Factor weights of selected indicators

Indicators	PC1	PC2	PC3	PC4	PC5
Area planted			0.48		
Animal traction	0.44				
Farm equipment			0.65		
Subsidized fertilizer				0.44	
Commercial fertilizer		0.42			
Herbicide			0.47		
Livestock ownership		0.53			
Ownership of Economic trees	–0.53	0.26			
Income from crop sales	0.38		0.28		
Insurance					0.67
Wage income					0.64
Membership of an association		0.58			
Household size	0.47				
Age				0.50	
Literacy				–0.63	

emphasizing off-farm income which is an important source of income complementing farm income in rural Nigeria. Agricultural insurance also captures the importance of insuring farmers against risk and the importance of risk-bearing agencies such as the National Food Reserve Agency. This adaptive capacity is quite weak in all the agro-ecological zones, and the weakness is most pronounced in the southeast.

8.5 CONCLUSION

Adaptive capacity is a complex concept and therefore indeterminate. Hence, several frameworks have emerged to operationalize adaptive capacity development assessment across sectors or regions. Based on Sustainable Livelihood and Entitlement Frameworks, policy-relevant indicators were isolated and weighted to give insight into the nature of farmers' adaptive capacity development in different zones in Nigeria. The analysis sheds light on the key structure and agro-input resources used by farmers across agro-ecological zones in response to the imperatives of the environment. The ability of farmers across all the zones to cope with and respond to problems of

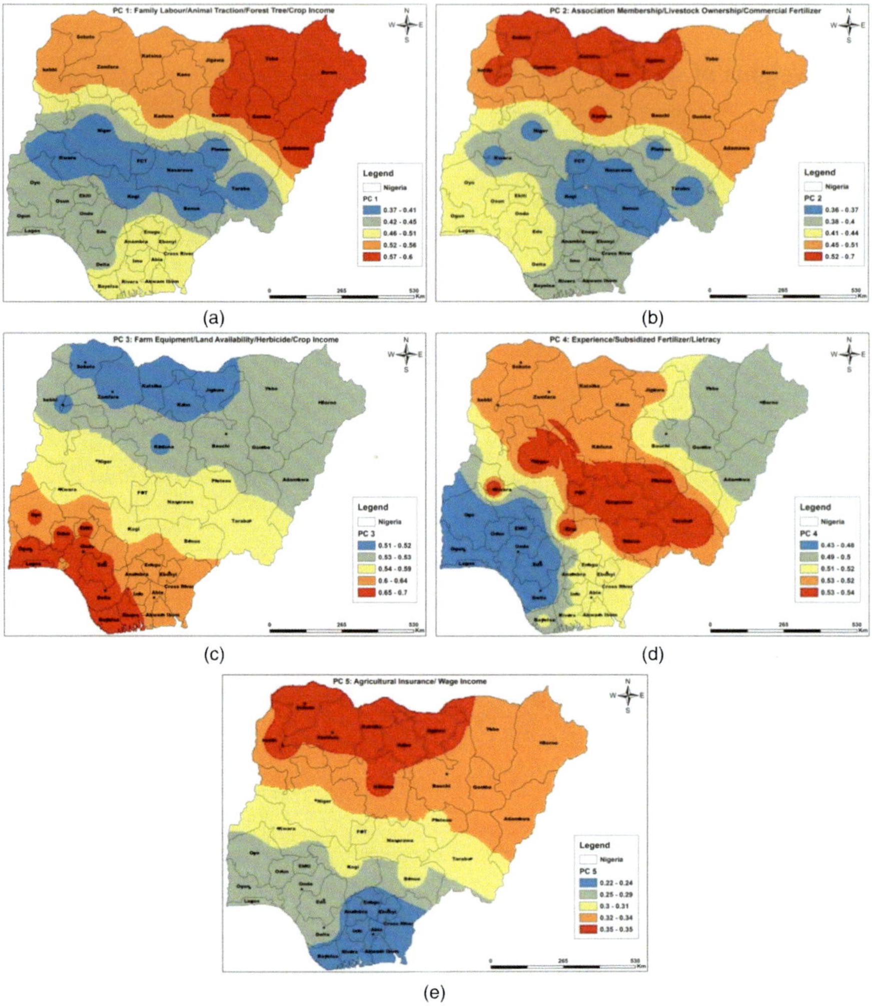

FIGURE 8.1A–E Maps highlighting results from Principal Components (PC1–PC5) across Nigeria's regions.

climate risk depends on access to these key inputs. Of importance is the weak use of agricultural insurance in all the agro-ecological zones. Hence the framework helps to raise questions on the underlying causes of weak adaptive capacity. The effectiveness of institutions both formal and informal is important in driving farmers' access to important agricultural inputs. Increasing adaptive capacity may require adaptation of both formal and informal institutions. Lack of availability and access to appropriate resources may significantly limit the ability of a system to cope with the effects of climate change and wider development pressures. The family labour and natural resources index was highest for states in Northeast and Northwest ecological zones with an index ranging from 0.52 to 0.6. The social and financial capital adaptive capacity index was highest for states in Northwest ecological zones ranging from 0.52 to 0.7.

The physical assets especially farm equipment and land index were most important in the southeast and southwest ecological zones. Human capital/access to subsidized fertilizer was most pronounced in all the states of Northern Nigeria. Finally agricultural insurance index was quite weak in all the agro-ecological zones and the weakness was most pronounced in southeast agro-ecological zone.

REFERENCES

Adger, W. N., & Barnett, J. (2009). Four reasons for concern about adaptation to climate change. *Environment and Planning A*, 41(12), 2800–2805.

Adger, W. N., Brooks, N., & Kelly, P. M. (2004). *New indicators of adaptive capacity*. Tyndall Centre for Climate Change Research. Available at: https://www.researchgate.net/publication/257343107_New_Indicators_of_Vulnerability_and_Adaptive_Capacity#fullTextFileContent

Asprone, D., & Manfredi, G. (2014). Linking disaster resilience and urban sustainability: A glocal approach for future cities. *Disasters*, 39(1), 96–111. https://doi.org/10.1111/disa.12106

Aulong, S., Chaudhuri, B., Farnier, L., Galab, S., Guerrin, J., Himanshu, H., & Prudhvikar, R. (2011). Are South Indian farmers adaptable to global change? A case in an Andhra Pradesh catchment basin. *Regional Environmental Change*, 12(3), 531–542. https://doi.org/10.1007/s10113-011-0258-1

Babulo, B., Muys, B., Nega, F., Tollens, E., Nyssen, J., Deckers, J., & Mathijs, E. (2008). Household livelihood strategies and forest dependence in the Highlands of Tigray, Northern Ethiopia. *Agricultural Systems*, 98(2), 147–155.

Barnes, M. L., Wang, P., Cinner, J. E., Graham, N. A., Guerrero, A. M., Jasny, L., & Zamborain-Mason, J. (2020). Social determinants of adaptive and transformative responses to climate change. *Nature Climate Change*, 10(9), 823–828.

Brooks, N., & Adger, W. N. (2004). *Assessing and enhancing adaptive capacity*: Technical Paper 7. UNDP. Available at: https://www.unisdr.org/files/7995_APF.pdf#page=170

Brooks, N., Adger, W. N., & Kelly, P. M. (2005). The determinants of vulnerability and adaptive capacity at the national level and the implications for adaptation. *Global Environmental Change*, 15(2), 151–163.

Campbell, B. M., Thornton, P., Zougmoré, R., van Asten, P., & Lipper, L., 2014. Sustainable intensification: What is its role in climate smart agriculture? *Current Opinion in Environmental Sustainability*, 8, 39–43.

Cinner, J. E., & Barnes, M. L. (2019). Social dimensions of resilience in social-ecological systems. *One Earth*, 1(1), 51–56.

Cinner, J. E., Adger, W. N., Allison, E. H., Barnes, M. L., Brown, K., Cohen, P. J., & Morrison, T. H. (2018). Building adaptive capacity to climate change in tropical coastal communities. *Nature Climate Change*, 8(2), 117–123.

Copley, N., Bowring, J., & Abbott, M. (2015). Thinking ahead: Design-directed research in a city which experienced fifty years of sea level change overnight. *Journal of Landscape Architecture*, 10(2), 70–81. https://doi.org/10.1080/18626033.2015.1058577

Decancq, K., & Lugo, M. A. (2013). Weights in multidimensional indices of well-being: An overview. *Weights in Multidimensional Indices of Wellbeing: An Overview. Econometric Reviews*, *32*(1), 7–34.

Dowd, A. M., Marshall, N., Fleming, A., Jakku, E., Gaillard, E., & Howden, M. (2014). The role of networks in transforming Australian agriculture. *Nature Climate Change*, 4, 558–563. https://doi.org/10.1038/nclimate2275

Engle, J. (2017). Stories of tragedy, trust and transformation? A case study of education-centred community development in post-earthquake Haiti. *Progress in Planning*, *124*, 1–34. https://doi.org/10.1016/j.progress.2017.04.001

Fazey, I., Wise, R. M., Lyon, C., Câmpeanu, C., Moug, M., Davies, E. T., & Anderies, J. M. (2015). Past and future adaptation pathways. *Climate and Development*, *8*(1), 26–44. https://doi.org/10.1080/17565529.2014.989192

Filmer, D., & Pritchett, L. H. (1998). *The effect of household wealth on educational attainment: Demographic and health survey evidence*. Policy Research Working Paper 1980. The World Bank.

Filmer, D., & Pritchett, L. H. (2001). Estimating wealth effects without expenditure data—Or tears: An application to educational enrollments in states of India. *Demography*, *38*(1), 115–132.

Gallopín, G. C. (2006). Linkages between vulnerability, resilience, and adaptive capacity. *Global Environmental Change*, 16(3), 293–303.

Garschagen, M. (2015). Risky change? Vietnam's urban flood risk governance between climate dynamics and transformation. *Pacific Affairs*, 88(3), 599–621.

Haddad, B. M. (2005). Ranking the adaptive capacity of nations to climate change when sociopolitical goals are explicit. *Global Environmental Change*, 15(2), 165–176.

Howden, S. M., Soussana, J. F., Tubiello, F. N., Chhetri, N., Dunlop, M., & Meinke, H. (2007). Adapting agriculture to climate change. *Proceedings of the National Academy of Sciences of the United States of America*, 104(50), 19691–19696.

IAASTD. (2009). *Agriculture at a crossroads: International assessment of agricultural knowledge, science, and technology for development* (Global Report). IAASTD.

IPCC. (2012). Managing the risks of extreme events and disasters to advance climate change adaptation. In C. B. Field et al. (Eds.) *A special report of working groups I and II of the intergovernmental panel on climate change* (vol. 3, pp. 25–64). Cambridge University Press.

IPCC. (2014). *Climate change 2014: Impacts, adaptation, and vulnerability. Part A: Global and sectoral aspects*. Cambridge University Press.

Jones, L., Ludi, E., & Levine, S. (2010). Towards a characterisation of adaptive capacity: A framework for analysing adaptive capacity at the local level. *ODI Background Note*, December 2010. Available at: https://media.odi.org/documents/6353.pdf

Kates, R. W., Travis, W. R., & Wilbanks, T. J. (2012). Transformational adaptation when incremental adaptations to climate change are insufficient. *Proceedings of the National Academy of Sciences of the United States of America*, 109(19), 7156–7161. https://doi.org/10.1073/pnas.1115521109

Kelly, P. M., & Adger, W.N. (2000). Theory and practice in assessing vulnerability to climate change and facilitating adaptation. *Climatic Change*, 47(4), 325–352.

Mapfumo, P., Onyango, M., Honkponou, S. K., El Mzouri, E. H., Githeko, A., Rabeharisoa, L., & Agrawal, A. (2014). Pathways to transformational change in the face of climate impacts: An analytical framework. *Climate and Development*, *9*(5), 439–451. https://doi.org/10.1080/17565529.2015.1040365

Marshall, N. A., Park, S. E., Adger, W. N., Brown, K., & Howden, S. M. (2012). Transformational capacity and the influence of place and identity. *Environmental Research Letters*, 7(3), 034022. https://doi.org/10.1088/1748-9326/7/3/034022

Marshall, N. A., Dowd, A. M., Fleming, A., Gambley, C., Howden, M., Jakku, E., & Thorburn, P. J. (2014). Transformational capacity in Australian peanut farmers for better climate adaptation. *Agronomy for Sustainable Development*, 34(3), 583–591. https://doi.org/10.1007/s13593-013-0186-1

Nelson, R., Kokic, P., Crimp, S., Meinke, H., & Howden, S. M., (2010). The vulnerability of Australian rural communities to climate variability and change: Part I—Conceptualising and measuring vulnerability. *Environmental Science & Policy*, 13(1), 8–17.

Park, S. E., Marshall, N. A., Jakku, E., Dowd, A. M., Howden, S. M., Mendham, E., & Fleming, A. (2012). Informing adaptation responses to climate change through theories of transformation. *Global Environmental Change*, 22(1), 115–126.

Reddy, V. R., Reddy, M. G., Galab, S., Soussan, J., & Springate-Baginski, O. (2004). Participatory watershed development in India: Can it sustain rural livelihoods? *Development and Change*, 35(2), 297–326.

Rickards, L., & Howden, S. M. (2012). Transformational adaptation: Agriculture and climate change. *Crop and Pasture Science*, 63(3), 240–250. https://doi.org/10.1071/CP11172

Smit, B., & Wandel, J. (2006). Adaptation, adaptive capacity and vulnerability. *Global Environmental Change*, 16(3), 282–292.

Vincent, K. (2007). Uncertainty in adaptive capacity and the importance of scale. *Global Environmental Change*, 17(1), 12–24.

Warner, B. P., Kuzdas, C., Yglesias, M. G., & Childers, D. L. (2015). Limits to adaptation to interacting global change risks among smallholder rice farmers in Northwest Costa Rica. *Global Environmental Change*, 30, 101–112.

Williges, K., Mechler, R., Bowyer, P., & Balkovic, J. (2016). Towards an assessment of adaptive capacity of the European agricultural sector to droughts. *Climate Services*, 1, 1–12.

World Economic Forum. (2001). *Environmental sustainability index*. Davos: World Economic Forum, Yale Center for Environmental Law and Policy & Center for International Earth Science Information Network.

Yohe, G., & Tol, R. S. J. (2002). Indicators for social and economic coping capacity – moving toward a working definition of adaptive capacity. *Global Environmental Change*, 12(1), 25–40.

Section IV

Cross-Cutting Issues

9 Climate-Resilient Health Systems for Sustainable Development

Işıl Demirtaş and Seda Yildirim

9.1 INTRODUCTION

Global warming and climate change cause many disasters such as extreme heat, floods, inundations, rising sea levels, and droughts. These natural disasters caused by climate change affect human health in many ways, both directly and indirectly. Every year, deaths occur in various parts of the world due to heat waves (Poynting and Stallard, 2024). Sustainable community health is one of the fundamental issues in building a sustainable future. The third fundamental goal within the 2030 SDGs includes the theme of health. (Yıldırım et al., 2020). The 2030 SDGs are essentially a guide to targets set to address 17 key global issues. In 2015, the 2030 SDGs, officially declared by the United Nations, have been accepted by UN member countries (Yıldırım et al., 2024a) and sustainable policies have begun to be developed to achieve the relevant goals (Yıldırım and Kantarcı, 2022; Yıldırım et al., 2024b).

As seen in Figure 9.1, the position of health issues among the 2030 SDGs and climate change relationships are shown.

According to studies and annual reports on the 2030 SDGs, significant progress has been made on some goals and countries are striving to work toward a sustainable future. However, the COVID-19 crisis that emerged in 2019 has caused chaos in many areas, especially in health (Bostancı et al., 2022; Öncü et al., 2021). The COVID-19 pandemic was accepted as a significant turning point that changed the

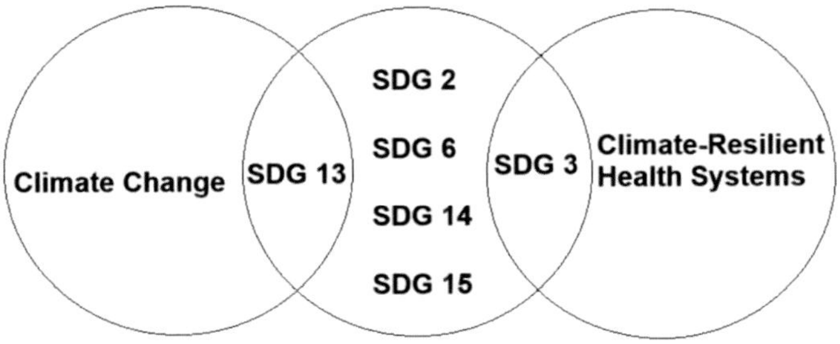

FIGURE 9.1 The position of health among 2030 SDGs.

Source: created by the authors.

DOI: 10.1201/9781003518037-13

entire world in the 21st century (Joseph and Kutty Narayanan, 2022). The COVID-19 pandemic has been a period that has proven that both developed and developing countries are inadequate in creating sustainable public health (Bostancı et al., 2023; Pollard et al., 2020). Therefore, it has shown the inadequacy of the health sector during the COVID-19 crisis and has led countries to create a resilient health system in unexpected outbreaks such as global virus outbreaks and pandemics (Cohen et al., 2020; United Nations, 2020; Liu et al., 2020; Olivia et al., 2020). At this point, the aim of this section is to discuss the necessity of establishing resilient health systems in countries and to make recommendations on establishing sustainable health systems within the scope of the 2030 SDGs. The COVID-19 pandemic may not have been successfully managed administratively. For example, there have been disruptions in patient treatment, economic contraction, and future problems for public health. Due to the COVID-19 pandemic, people's primary source of information is social media and the internet, so access to incorrect data on health has increased (Luqmani and El Hashim, 2022).

During the COVID-19 pandemic, all individuals and institutions globally have tried to survive with the practices and information based on the World Health Organization (WHO). WHO has become an institution that all world leaders and individuals focus on during the COVID-19 pandemic and has become the main source of information to get rid of the pandemic. WHO has provided a wide range of information systems related to COVID-19 as open access that every user can access (WHO, 2023a).

It has been argued that the COVID-19 pandemic has also emerged because of climate change, with an indirect effect. The COVID-19 pandemic has shown that humanity is not prepared and equipped enough for a global health crisis in the 21st century (Blair, 2020; OECD, 2023). Therefore, it has been seen that health policies and the health system should be planned in a way that will achieve sustainable public health for a sustainable future.

Health issues and climate change should be handled together in the long term. WHO's COP 29 Climate Change and Health Special Report titled "Health is the Argument for Climate Action" emphasized that health should be at the center of climate. Although solutions to climate change will reduce health problems, focusing solely on climate change will not be enough to address all dimensions of this impact. Therefore, health and climate should be considered together. The WHO emphasizes that combating climate change should be addressed not only to prevent environmental damage, but also because it threatens public health, and states that health is an important source of motivation in this fight (WHO, 2024).

Although the close relationship between climate change and health has been addressed in previous COP meetings, the Alliance for Transformative Action on Climate and Health (ATACH) was established by WHO at COP26. ATACH was established to achieve the goal of establishing climate-resilient and sustainable health systems determined at COP26. In line with this goal, the Climate Resilient Health Systems working group was established (WHO, 2022a).

Climate-resilient health system is one of the issues that require health and climate change to be addressed holistically. Although weather events such as drought, flood, sea level rise, and extreme temperatures can be predicted with today's developing technologies, it is often not possible to predict the damage they can cause

(Campbell-Lendrum et al., 2023; Mosadeghrad et al., 2023; ClearIAS Team, 2023). For this reason, sudden shocks make health services and health infrastructure fragile. Therefore, the resilience of the health system is important in terms of reducing the degree of exposure to global warming and climate change. This study will examine the relationship between health issues and climate change in the long term. Due to the vulnerabilities created by climate change in the health sector, the necessity of establishing climate-resilient health systems will be emphasized. In addition, it will be evaluated that addressing climate change and sustainable development goals together will play an important role in establishing new and alternative policies to set climate-resilient health models or systems

9.2 CLIMATE CHANGE-BASED HEALTH RISKS

Making health systems climate resilient is very important in terms of eliminating and preventing climate-related health risks. Accordingly, it is important to identify the possible risks that health systems may face (WHO, n.d.a). Climate change causes different natural events. The sixth report of IPCC stated that natural events caused by climate change were expressed in six categories as flooding, droughts, heat waves, tropical storms, wildfires, sea level rise. These categories can be seen as vital and global issues that threaten human health such as water, food inadequacy, and access to clean water. In addition, extreme heat also brings diseases such as metabolic and circulatory system diseases (Palmeiro et al., 2023; Schwerdtle et al., 2024). When considering the direct or indirect influence of climate change on health issues, IPCC gives three categories as "direct effects, influences originating from natural systems and effects related to the human system, and health risk outcomes" as seen in Table 9.1. Effects such as extreme heat, extreme humidity, and extreme increases in water temperatures are the results of climate change, and it has been shown that these weather events cause a series of health problems. These problems bring many health risks including death, metabolic and circulatory system diseases, and infectious diseases (Handmer et al., 2012).

Direct effects generally occur due to the higher temperature. Chronic diseases and deaths can occur directly on human health due to temperature changes or excessive heat. Secondly, negative impacts caused by natural systems may cause water and food shortages related to sea level temperature, freshwater temperature, and humidity factors. Low- and middle-income countries and vulnerable groups (children, the elderly, the poor, and the sick) where access to clean water and sufficient food is difficult are particularly affected (WHO, 2015). On the other hand, water shortage and respiratory system-related diseases are also long-term negative outcomes of climate change on people. Climate-related disasters can also have impacts on people's mental health. Climate disasters cause people to lose their loved ones, to be displaced, and to experience mental illnesses such as stress and anxiety due to seasonal changes. On the other hand, the influences of deteriorating air quality on respiration can also affect human life in the long term (World Economic Forum, 2024). Seasonal temperature changes depending on average temperature changes are other influences that lead to illnesses and deaths. Weather events such as being hotter than average in summer and colder than average in winter can have different problems on the heat balance of the human body (USGCRP, 2016).

TABLE 9.1
Climate change, health risks and outcomes

Climate change effects	Causes of health risks	Health risk outcomes
Direct effects	Extreme weather	Deaths due to extreme heat, heat exhaustion, and heat stroke, metabolic and circulatory system diseases
Effects related to the natural system	Extreme weather, high humidity, increased precipitation, rising sea levels, and rising freshwater temperatures	Water, foodborne, and vector-borne diseases due to changes in pathogenic activity, ecological and seasonal variations, and flood-related damage to water and sanitation infrastructure
Effects related to human system	Extreme weather, precipitation, and humidity changes	Malnutrition, infectious diseases due to insufficient food production in the tropics, stunting, and wasting in children

Source: adapted from WHO (2015).

Air pollution, like heat, is a significant threat to human health. According to the WHO (2013) statistics, it was seen that Africa and Southeast Asia were the regions with the highest loss of life years due to climate change. In these regions, problems such as food insufficiency, access to clean water, and the presence of disadvantaged groups, as well as the inability to prevent climate change, are increasing the problems in the region (WHO, 2013). In addition, of the 387 disasters that occurred globally in 2022, 79 occurred in Africa (WHO and IAHO, 2024).

9.3 THE NEGATIVE ROLE OF CLIMATE CHANGE ON HEALTH SYSTEMS AND THE RELATIONSHIP WITH SDG TARGETS

Natural disasters such as drought, flood, and hurricane caused by climate change (Bolan et al., 2024; Poynting and Stallard, 2024; Furtak and Wolińska, 2023) can suddenly and unexpectedly increase the need for health systems. The more severe the impact of natural disasters, the greater the need for the health system. Although the timing and severity of natural disasters can be predicted with the developed early warning systems, it is sometimes quite difficult to estimate the extent of the damage that will be caused. On the other hand, it is also quite difficult to predict how much demand will be created on the health system by epidemics triggered by disasters, such as access to clean water and food. Other problems that natural disasters will create are also related to the effectiveness of health services. Problems that may occur in the infrastructure of health service institutions due to disasters – such as energy and water outages - directly affect the effectiveness of health services – such as damage to the infrastructure of hospitals. In addition, the lack of sufficient health personnel or their inadequacy in meeting the demand also causes serious problems in the health system (Salam et al., 2023; Ebi et al., 2021). These problems with the health system reduce the capacity of the health system to respond to problems. Health services are disrupted and services related to the functioning of the health system, such as transportation and distribution, cannot be provided on time, and inequalities in access to health services increase (Deloitte, 2024). Impacts such as drought, flood, sea level rise, and high temperatures caused by climate change affect vulnerable groups more and even pose threats by damaging the resource and support infrastructure in the African region, where these groups are widespread. The fact that low-income communities live in hard-to-reach and climate-sensitive areas and cannot benefit from adequate health services further increases the influence of climate change on low-income communities (World Economic Forum, 2024). Along with low-income families, the elderly, children, women, and migrants are also vulnerable groups exposed to climate change. Particularly, those living at low-income levels are the most vulnerable segments of society (World Bank, 2024). The negative impacts or outcomes of climate change become more evident in countries with unplanned urbanization, environmental degradation, poverty, and poor governance (UN, 2012).

Climate change poses a serious challenge to the sustainability of human health by affecting food security, and the responsibility of health services is also increasing. The impacts caused by extreme climate events affect all dimensions of food security, including food accessibility, nutritional deficiencies, and food system resilience (El Bilali et al., 2020). The prevalence of malnutrition worldwide increased from 8.5% in 2020 to 9.1% in 2023. At this rate, it reached 20.4% in 2023 from 18.8% in the African region. It was 30.8% in 2023, especially in Central Africa (FAO et al., 2024). The Global Food Crises 2024 Interim Report states that malnutrition increased from 705,000 in 2023 to 1.9 million in 2024.

SDG 2 aims to eliminate hunger by 2030. The United Nations attributes the increase in hunger levels to many reasons such as conflicts, climate shocks, increases in food prices, and decreases in agricultural production. Climate shocks are quite remarkable for these reasons. They also affect the decrease in agricultural land and the increase in food prices. The negative outcomes of climate change make it difficult to achieve SDG 2.1 and 2.2, which are the 2030 Development Goals to end hunger and malnutrition (World Bank, 2024). Climate change also poses a major risk to access to clean water. Water, which is of vital importance to people, requires access to sufficient and healthy conditions to reduce diseases and maintain public health (WHO et al., 2022).

One of the negative outcomes of climate change is giving additional costs on health systems. According to the World Bank Group (2024), it is estimated that by 2050, there will be an excess cost of $21 trillion, equivalent to 1.3% of GDP in low- and middle-income countries. For example, 'hurricane Ivan' in the year 2004, cost Grenada 200% of GDP per capita, and the earthquake in Haiti in 2010 had an impact of 15% of GDP. For LDC economies, Bangladesh and Mozambique lost 3–5% of their GDP due to disasters (UN, 2012).

Protection from climate change will help achieve SDG targets. Thus, the climate-related vulnerability of the health sector can be prevented (Table 9.1). However, climate-related resilience also requires the health system to be prepared against climate impacts and risks. In other words, the health system should be organized by considering all these impacts and risks. While climate-resilient health systems are affected by external factors, this resilience also depends on institutional and infrastructural factors. Climate-resilient health systems and their characteristics are discussed in the below section.

9.4 THE CONCEPT OF CLIMATE-RESILIENT HEALTH SYSTEMS AND BUILDING BLOCKS

Climate change threatens human health by causing many health problems. The health system's readiness for shocks, its ability to absorb them, and its capacity to manage these capabilities are also effective in minimizing the impacts of climate change on health (Thomas et al., 2020). The ability of the health system to minimize the influences of climate change on human health depends on the development of climate-resilient health systems (IPCC, 2018). The COVID-19 Pandemic has played a major role in revealing and experiencing the importance of the resilience of health systems. With this crisis, it has become important to consider factors that will reduce fragility, such as preparing for, adapting to, and anticipating problems experienced in the health systems of countries (Sagan et al., 2021; Ledesma et al., 2024).

Climate change challenges the capacity of health systems. The combined and cascading factors of shocks on the health system require capabilities such as (i) developing adequate capacity and resilience; (ii) anticipating health risks; and (iii) responding to needs in a timely manner (WHO, 2015). For health systems to become climate resilient, it is important to reduce the vulnerabilities that will affect the health system. WHO has created an operational framework to guide member countries in establishing climate-resilient health systems. WHO (2015) recommends the development of six main components and ten subcomponents to build climate-resilient health systems (Figure 9.2). Launching climate-resilient health systems requires that each of the interrelated yet independent building blocks be climate resilient (WHO, 2015).

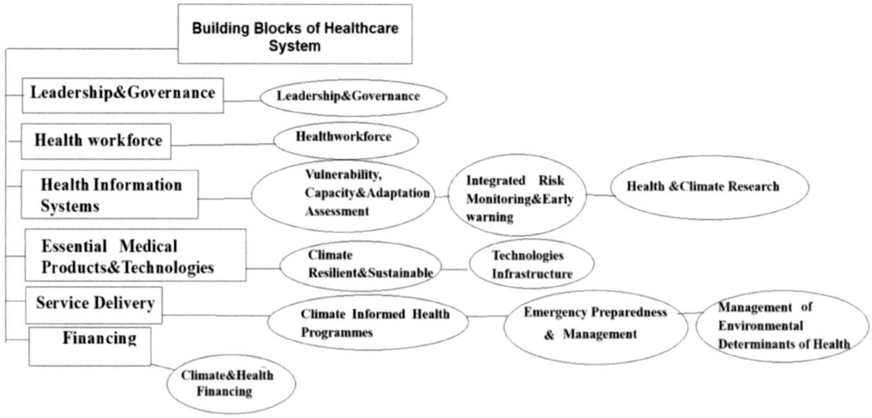

FIGURE 9.2 Model of climate-resilient health systems.

Source: adapted from WHO (2022b).

Based on the model of climate-resilience health systems as suggested by WHO (2022b), main concepts can be examined as follows.

9.4.1 Leadership and Governance

The World Health Organization emphasized in The World Health Report published in 2000 that governments have responsibilities in the health system, such as defining, organizing, collecting, and using information. Later, the World Health Organization expanded the scope by considering the concept of governance together with leadership (Smith et al., 2012). Making the health system climate resilient requires health policies to be designed by considering climatic stress and shocks (WHO, 2015). The evaluation of climate change within the scope of health policies is related to good governance. The following criteria are decisive for good governance (Kapologwe et al., 2023):

- **Participation:** Active participation of all stakeholders in the decision-making process.
- **Transparency:** Openness in sharing, using, understanding, and monitoring information at all levels of the health system.
- **Accountability:** Health sector employees are accountable to all stakeholders.
- **Responsiveness:** Sensitivity of health services to the needs of health users, meeting their expectations.
- **Health equity**: Ensuring access to health services for all social groups and meeting the needs of groups at the highest health risk.
- **Effectiveness and efficiency:** Optimal use of organizational resources.
- **Rule of law:** Impartiality of health policies, rules, institutions, and laws.
- **Ethics:** Existence and application of health ethics principles in health services.

9.4.2 Health Workforce

The health workforce is one of the most important components of health systems. A system with a qualified and sufficient workforce that responds to people's needs and expectations will play an important role in sustaining health policies (WHO, n.d.b).

In order for the health workforce to be used efficiently by policy makers, it is necessary to make accurate estimates regarding the availability of health workers at a level that will respond to public health services and take into account the needs of the population, and to consider factors such as the training of qualified health workers, employment, remuneration, distribution of the workforce and financing (WHO, 2015). Although different elements of the health workforce stand out among countries, several common requirements can be listed to increase the effectiveness of health personnel (WHO, 2010):

- Adequate workforce
- Effective functioning of the incentive system

- Effective distribution
- Establishment of ethical values and norms
- Existence of support systems
- Cooperation among stakeholders

Health workers are also at the forefront of resilient health systems. The balance between quality, demand, and supply of health workers directly affects the resilience of the system. Universal Health Coverage (UHC), defined under SDG 3, means that everyone has access to the health services they need. UHC requires closing the global health gap and ensuring equitable access to the necessary health personnel. To make progress toward UHC, closing the global health worker gap, thus ensuring that there are the right number of adequate health workers in the right places in line with the health needs of the population, is of critical importance. Table 9.2 shows the density of health workers by region. It is seen that African countries have a very low density of health workers compared to other regions. The countries with a high density are developed countries (WHO, 2023b).

The preparedness of health workers for climate change conditions is considered by the WHO as a fundamental component of the resilience of health systems. Understanding, analyzing, predicting climatic conditions, and being prepared in case of risk and shock to ensure sound decision-making and managing the problems related to climate change are the requirements of a well-functioning health system. Effective management of changing conditions depends on health workers' human resources, development of organizational capacity, and communication and awareness (WHO, 2015).

9.4.3 Health Information Systems

Health Information Systems includes three main concepts as (i) Vulnerability, Capacity, and Adaptation Assessment, (ii) Integrated Risk Monitoring and Early Warning, and (iii) Health and Climate Research.

i. Vulnerability, Capacity and Adaptation Assessment: The risk element is at the center of the Vulnerability, Capacity and Adaptation Assessment and defined as a tendency (WHO, 2012). Health risks vary depending on whether the population is in a risky area in terms of climate change, the structure of socioeconomic groups, the problems caused by climatic changes, and the capacity of the health system. In addition, climate change does not affect all segments of society equally, so it also increases existing health inequalities (Deloitte, 2024). WHO (2012) explained the negative health outcomes of climate change depending on sociopolitical, biological, socioeconomic, geographical, and demographic conditions. Establishing an adjustable health policy mechanism for adaptation in health systems improves the ability of communities to adapt to climate change and minimize risks (WHO, 2012). The development of the adaptation mechanism is possible by determining the risk factors locally and adapting health policies to be prepared for these risks.

TABLE 9.2
Distribution of healthcare workers density by region (per 10,000 population)

Areas	Medical doctors	Nursing and midwifery personnel	Dentists	Pharmacists
African region	2,6	11,3	0,2	0,4
Region of the Americas	28,0	67,1	5,5	6,2
South-East Asia region	7,7	20,6	1,5	6,4
European region	37,6	82,6	6,3	6,7
Eastern Mediterranean region	11,6	16,6	2,9	3,5
Western Pacific region	23,2	43,4	4,6	4,7
Global	17,2	37,7	3,3	4,8

Source: adapted from WHO, (n.d.c).

ii. Integrated Risk Monitoring and Early Warning: It includes the prediction, monitoring, and establishment of early warning systems for high risks. Early warning systems enable the prediction of environmental health risk and the preparation of health personnel and the public to take the necessary precautions in the process (WHO, 2015).

According to the United Nations Office for Disaster Risk Reduction (UNDRR, 2025), there are four main components of early warning systems:

- Collecting data to identify risks
- Identifying the possible consequences of risks
- Ensuring timely and accurate dissemination of information
- Preparing for the response system

The success of early warning systems will ensure both the efficient use of available resources and the early action taken against environmental risks, preventing the risk from growing and affecting fewer people. Thus, it will contribute to the resilience of the health system.

iii. Health and Climate Research: The development of the resilience of the health system is directly related to scientific studies conducted in the field of climate and health. The World Bank has determined that one of the three main pillars of the climate and health program is the creation of evidence and knowledge. The three components of the climate and health program recommended by the World Bank for countries are given in Table 9.3.

TABLE 9.3
Components of the climate and health program

Create evidence and knowledge	Investment in country-specific evidence	Partnership
• Detection of vulnerabilities related to climate risks • Considering impacts on communities and economies • Determine investment strategies to develop, finance, and measure evidence	• Investment in low-carbon resilient health systems • Leverage the World Bank's health incentives • Investments in early warning systems, improving health services, climate-driven diseases, facilities, and health workers	• Collaboration • Actively participate in Development Banks' work on Climate-Health Finance • Private sector partnerships for capital

Source: adapted from World Bank, (n.d.).

Obtaining information about the health consequences of climate-related risks and using this information to generate strong evidence through applied research will increase the climate resilience of health systems (WHO, 2015).

9.4.4 Essential Medical Products and Technologies

Climate resilient – sustainable technologies and infrastructure part is related with essential medical products and technologies due to achieving a health resilience system. The provision of health equipment is necessary for a climate-resistant and sustainable health system. This equipment should include both preventive and curative health equipment. On the other hand, the existence of technologies that meet the needs of the health system is necessary, and the existence of technologies that predict climate risks outside the health system is directly related to the resilience and sustainability of the health system (WHO, 2015). Building infrastructure that is climate-resistant will prevent disruptions to healthcare services. Infrastructure services are the services most exposed to climate risks. For example, building energy, water, and sewage infrastructure that is climate-resistant will prevent disruptions to these services that are directly related to healthcare services (OECD, 2024).

9.4.5 Service Delivery

In this section, three main issues will be examined as follows:

i. **Management of Environmental Determinants of Health:** Health systems need effective management that includes planning, predicting, coordinating, and regulating the damaging issues of health systems on the environment. Environmental determinants related to health management are listed below (WHO, 2020):

- Access to sufficient and clean water is necessary for health systems to be provided in hygienic conditions.
- Health wastes must be stored and disposed of in appropriate conditions.
- Wastewater treatment and sanitation systems must be established to prevent infectious diseases.
- Health facilities use chemicals that pose environmental and health risks. Necessary precautions must be taken in the use of these chemicals.
- Necessary precautions must be taken in the use of radiation in the treatment of diseases.
- Air quality: Providing the energy needs of health facilities from renewable energy instead of fossil fuels.
- Food: Making food facilities in health facilities environmentally friendly

Ensuring that the health system can provide both the infrastructural and service supply needs required for its efficient operation and the ability to provide services in harmony with the environment is possible by managing health management with environmental awareness. Health systems need to develop policies in an inclusive manner, integrating all systems affected by climate (WHO, 2015).

ii. **Climate Informed Health Programs:** Health programs are needed for the health sector to respond to climate-related risks. These programs are essential for predicting climate change, determining the measures to be taken, coordination, and planning (WHO, 2015).

iii. **Emergency Preparedness and Management:** Being prepared for health risks is referred to as emergency preparedness. WHO emphasizes the need for all organizations and governments to follow an emergency cycle to reduce the impact of disasters. This cycle consists of prevention, preparedness, response, and recovery stages (WHO, n.d.d):
   - Prevention: Detecting health problems before they occur and preventing them from occurring. Although it is not possible to prevent natural disasters, the health problems that these disasters will cause can be prevented or minimized.
   - Preparedness: The entire health system is ready to manage risks. It constitutes the process of conducting a strategic risk assessment and taking the necessary measures according to this assessment.
   - Respond: It is the implementation of necessary interventions with effective emergency response.
   - Recovery: It is the renewal of the system in a way that considers post-action reviews and lessons learned.

WHO recommends an Incident Management System (IMS) for emergencies. The components of this system are as follows and illustrated in Figure 9.3.

These components are necessary for a rapid, coordinated, and efficient response. The strategic decision-making ability and coordination skills of the management unit that make critical decisions are the primary criteria for the smooth functioning of the entire organization.

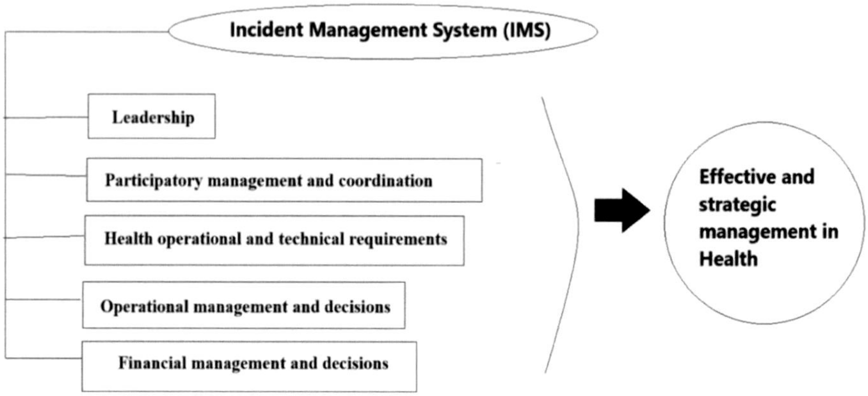

FIGURE 9.3 Key Components of the Incident Management System (IMS).

Source: created by the authors based on WHO guidelines, 2017.

9.4.6 Financing

In part of financing, Climate and Health Financing is related to the Healthcare Resilience System. To cope with and reduce the negative influence of climate change on healthcare sector and society's health, there is a need for an efficient allocation of financial resources. A health system that is organized to respond to the impacts of climate change must also be financially prepared in the event of a possible shock. Public resources and private sector investments also play a role in establishing climate-resilient health systems. Governments can allocate budgets to the health system or encourage private sector investments to eliminate the negative sides of climate change (WHO, 2023b).

9.5 CARBON NEUTRALITY AND HEALTH SYSTEMS

The healthcare sector presents contradictions in its relationship with climate change. On the one hand, it provides services aimed at improving human health, while on the other hand, it contributes to environmental pollution through the services it delivers. The Net Zero Emissions Strategy requires the healthcare sector to set priorities regarding emission reductions. It is known that the healthcare sector lags other service sectors in reducing emissions. The operation of large hospitals, high energy consumption, and waste generation increase the sector's contribution to environmental pollution. Depending on the country, the sector is responsible for approximately 4–8.5% of carbon emissions (Braithwaite et al., 2024).

In addition, due to the strategic importance of the healthcare sector, its economic scale, its responsibility to improve human health, and its high carbon footprint, it is necessary for healthcare institutions to develop specific carbon reduction policies. Strategies particularly focusing on harm reduction and adaptation have come to the forefront (Sladek et al., 2024). The healthcare sector must develop strategies to reduce its carbon footprint and achieve a sustainable health system.

First and foremost, using sustainable materials instead of traditional ones is essential. Replacing environmentally harmful materials such as plastic gloves, plastic IV bags, and plastic cutlery with eco-friendly medical materials is a fundamental step to be taken (Firmo, 2024). Furthermore, materials used in surgical operations contribute significantly to greenhouse gas emissions. Strategies such as reducing the number of surgical products used, ensuring their repair, and enabling their reuse are efficient practices for reducing emissions (Rizan et al., 2023). In addition, logistics services used to optimize healthcare delivery are not always aligned with emission reduction goals. Therefore, while establishing supply chains, it is necessary to consider carbon emissions alongside cost and efficiency. Relevant regulations on healthcare logistics should be shaped accordingly (Pohjosenperä et al., 2022). Digital health is also one of the prominent approaches to reducing carbon emissions. Telemedicine, which enables healthcare services to be delivered more efficiently by overcoming barriers such as time and distance through information and communication technologies, is suggested as a tool for reducing the ecological footprint (Holmner et al., 2014). Developments in information and communication technologies that allow services such as appointment scheduling and accessing test results to be provided digitally can contribute significantly to reducing the carbon footprint of healthcare services (Morcillo Serra et al., 2023).

The significant share of emissions originating from the healthcare sector indicates that carbon reduction policies should also be prioritized in the development of health policies. Another important aspect that needs to be considered in these policies is the ecological footprint caused by the transportation of patients and healthcare staff to medical facilities. The United Kingdom's National Health Service (NHS) states that the transportation of patients, visitors, and staff accounts for approximately 10% of total healthcare service emissions (Zurl et al., 2024).

9.6 CONCLUSION

Climate change poses serious threats to human health, and the extent of these threats depends on both the steps to be taken to prevent climate change and the resilience of health systems and their capacity to cope with these impacts. In this context, the success of climate-resilient health systems depends on the implementation of internal factors aimed at reducing climate change and climate-related impacts, and external factors such as institutional arrangements that will make the health system infrastructure resilient. A climate-resilient health sector is also compatible with SDG targets. Achieving SDG targets will play an important role in mitigating the impact of climate change on the health sector. SDGs related to the health impacts of climate change, health risks associated with these SDGs, and policies to be implemented to reduce these risks are summarized in Table 9.4.

Achieving SDG targets will also reduce the pressure on the health system. However, of course, this is not enough. Because the ability of the health system to eliminate risks is also related to how health systems are organized. Health systems should be made climate-resistant and their fragility should be reduced. For this, the ten components put forward by the WHO play a critical role. Components such as the institutional infrastructure, organization, manpower, and financing opportunities

TABLE 9.4
Policy recommendation for SDGs and health issues

SDGs	Negative elements based on climate change	Policy recommendations for climate change and health issues
SDG 2: Zero Hunger	Hunger, infectious diseases, food security, and water stress, psychological effects	Promoting sustainable agricultural practices, protecting from the negative sides of climate change, preventing food waste, and supporting local agriculture
SDG 3: Good Health and Well-Being	Infectious diseases, respiratory diseases, water and food security, psychological effects	Preventive health services, providing access to health services for everyone, strengthening health infrastructure
SDG 6: Clean Water and Sanitation	Infectious diseases, hygiene problems, food security due to inadequate irrigation of agricultural areas, diseases caused by pollution of water resources	Improving sanitation infrastructure, protecting clean water resources, reducing water waste, efficiency of water distribution system
SDG 13: Climate Action	Infectious diseases, deaths, problems of access to clean food and water	Transition to renewable energy, protecting forests and afforestation, sustainable transportation, sustainable agriculture, energy efficiency, waste management
SDG 14: Life Below Water	Seafood insecurity	Fighting against pollution to protect seas and oceans, protecting marine ecosystems and biodiversity, making fishing activities sustainable
SDG 15: Life on Land	Infectious diseases, deaths, problems of access to clean food and water	Increasing forest protection and afforestation efforts, implementing sustainable land management techniques, protecting biodiversity,

Source: created by the authors

of the health system will be effective in reducing the fragility of the health system by ensuring that the health sector operates as a resilient and flexible system.

Many natural disasters such as floods, droughts, extreme weather events, and sea level rise resulting from climate change and global warming pose risks that threaten human health. Infectious diseases, respiratory diseases, and deaths occur due to extreme heat. On the other hand, access to clean water and food inadequacy can also have an impact on human health because of disasters. Health risks are also directly related to SDG 2, 3, 6, 13, 14, and 15 targets in Table 9.4. Droughts caused by climate change affect people's access to food in some regions, causing health problems due to malnutrition. On the other hand, damage to infrastructure because of floods creates problems in accessing clean water, leading to hygiene and thirst problems and increasing infectious diseases. Therefore, to achieve SDG targets, the negative sides of climate change must be reduced, and appropriate policies must be implemented to achieve the targets. Achieving the targets will reduce climate-related health problems and alleviate the fragility of the health sector.

The health sector needs to be made climate resilient, both by addressing the health sector's own institutional structure and internal dynamics, and by addressing it together with the SDG targets. The study in question emphasizes the necessity of climate-resilient health systems and the importance of SDG targets in increasing the resilience of these systems. However, it emphasizes that the health sector's climate resilient should be evaluated holistically, not only with the SDG targets related to Climate Change (SDG 13) and the health sector (SDG 3).

REFERENCES

Blair, T. (2020). *The World Was Not Prepared for Covid-19. We Should Learn the Lessons and Change*. https://institute.global/insights/public-services/world-was-not-prepared-covid-19-we-should-learn-lessons-and-change

Bolan, S., Padhye, L. P., Jasemizad, T., Karmegam, M. G. N., Wijesekara, H., Amarasiri, D., Hou, D., Zhou, P., Biswal, B. K., Balasubramanian, R., Wang, H., Siddique, K. H. M., Rinklebe, J., Kirkham, M. B., & Bolan, N. (2024). Impacts of climate change on the fate of contaminants through extreme weather events. *Science of the Total Environment*, 909, 168388. https://doi.org/10.1016/j.scitotenv.2023.168388

Bostancı, S. H., Yildirim, S., & Erdogan, F. (2022). A review on e-government portal's services within hospital information system during Covid-19 pandemic. *Konuralp Medical Journal*, 14(S1), 271–279.

Bostancı, S. H., Yıldırım, S., & Yildirim, D. C. (2023). A study on next-generation digital tool for health data management: The e-Pulse portal. *International Journal of Health Governance*, 28(3), 238–248. https://doi.org/10.1108/IJHG-09-2022-0082

Braithwaite, J., Smith, C. L., Leask, E., Wijekulasuriya, S., Brooke-Cowden, K., Fisher, G., Patel, R., Pagano, L., Rahimi-Ardabili, H., Spanos, S., Rojas, C., Partington, A., McQuillan, E., Dammery, G., Carrigan, A., Ehrenfeld, L., Coiera, E., Westbrook, J., & Zurynski, Y. (2024). Strategies and tactics to reduce the impact of healthcare on climate change: Systematic review. *BMJ,* 387, e081284. https://doi.org/10.1136/bmj-2024-081284

Campbell-Lendrum, D., Neville, T., Schweizer, C., & Neira, M. (2023). Climate change and health: Three grand challenges. *Nat Med*, 29, 1631–1638. https://doi.org/10.1038/s41591-023-02438-w

ClearIAS Team (2023). *Climate Resilient Health Systems*. https://www.clearias.com/climate-resilient-health-systems/

Cohen, S. P., Baber, Z. B., Buvanendran, A., McLean, B. C., Chen, Y., Hooten, W. M., Laker, S. R., Wasan, A. D., Kennedy, D. J., Sandbrink, F., King, S. A., Fowler, I. M., Stojanovic, M. P., Hayek, S. M., & Phillips, C. R. (2020). Pain management best practices from multispecialty organizations during the COVID-19 pandemic and public health crises. *Pain Medicine*, 21(7, July 2020), 1331–1346. https://doi.org/10.1093/pm/pnaa127

Deloitte, Project Hope (2024). *Empowering the health care workforce for a climate-resilient future*.

Ebi, K. L., Vanos, J., Baldwin, J. W., Bell, J. E., Hondula, D. M., Errett, N. A., Hayes, K., Reid, C. E., Saha, S., Spector, J., & Berry, P. (2021 Apr). Extreme weather and climate change: Population health and health system implications. *Annual Review of Public Health, 1*;42, 293–315. https://doi.org/10.1146/annurev-publhealth-012420–105026. Epub 2021 Jan 6. PMID: 33406378; PMCID: PMC9013542.

El Bilali, H., Bassolé, I., Dambo, L., & Berjan, S. (2020). Climate change and food security. *Agriculture and Forestry*, 66, 197–210. https://doi.org/10.17707/AgricultForest.66.3.16

FAO, IFAD, UNICEF, WFP and WHO. (2024). *The State of Food Security and Nutrition in the World 2024 – Financing to end hunger, food insecurity and malnutrition in all its forms*.

Firmo, A. (2024, Nov 25). Reducing hospital carbon footprint: 20 sustainable alternatives to traditional medical supplies. *NetNada*. https://www.netnada.com/post/reducing-hospital-carbon-footprint-20-sustainable-alternatives-to-traditional-medical-supplies

Furtak, K., & Wolińska, A. (2023). The impact of extreme weather events as a consequence of climate change on the soil moisture and on the quality of the soil environment and agriculture – A review. *CATENA*, 231, 107378. https://doi.org/10.1016/j.catena.2023.107378

Governance and Development (English). The World Bank. https://documents.worldbank.org/curated/en/604951468739447676/Governance-and-development

Handmer, J., Honda, Y., Kundzewicz, Z. W., Arnell, N., Benito, G., Hatfield, J., Mohamed, I. F., Peduzzi, P., Wu, S., Sherstyukov, B., Takahashi, K., & Yan, Z. (2012). 4 changes in impacts of climate extremes: Human systems and ecosystem. In C. B. Field, V. Barros, T. F. Stocker, D. Qin, D. J. Dokken, K. L. Ebi, M. D. Mastrandrea, K. J. Mach, G. K., Plattner, S. K. Alien, M. Tignor, & P. M. Midgley (Eds.), *Managing the risks of extreme events and disasters to advance climate change adaptation*. (pp. 231–290). A Special Report of Working Groups I and II of the Intergovernmental Panel on Climate Change (IPCC). Cambridge University Press.

Holmner, Å., Ebi, K. L., Lazuardi, L., & Nilsson, M. (2014). Carbon footprint of telemedicine solutions – Unexplored opportunity for reducing carbon emissions in the health sector. *Journal of Cleaner Production*, 60, 492–501. https://doi.org/10.1016/j.jclepro.2013.04.019

IPCC (2018). Annex I: Glossary [Matthews, J. B. R. (ed.)]. In V. Masson-Delmotte, P. Zhai, H.-O. Pörtner, D. Roberts, J. Skea, P. R. Shukla, A. Pirani, W. Moufouma-Okia, C. Péan, R. Pidcock, S. Connors, J. B. R. Matthews, Y. Chen, X. Zhou, M. I. Gomis, E. Lonnoy, T. Maycock, M. Tignor, & T. Waterfield (Eds.), *Global Warming of 1.5°C. An IPCC Special Report on the impacts of global warming of 1.5°C above pre-industrial levels and related global greenhouse gas emission pathways, in the context of strengthening the global response to the threat of climate change, sustainable development, and efforts to eradicate poverty* (pp. 541–562). Cambridge University Press. https://doi.org/10.1017/9781009157940.008

Joseph, S., & Kutty Narayanan, A. (2022 Jan). Covid 19-the 21st century pandemic: The novel coronavirus outbreak and the treatment strategies. *Advanced Pharmaceutical Bulletin*, 12(1), 34–44. https://doi.org/10.34172/apb.2022.005. Epub 2021 Jan 30. PMID: 35517888; PMCID: PMC9012918

Kapologwe, N. A., Kalolo, A., Kengia, J., Swai, I. L., Kesale, A. M., Kacholi, G., ... & Anasel, M. G. (2023). Community health systems management and governance. In *Leadership and Governance in Primary Healthcare* (pp. 91–106). CRC Press.

Ledesma, J. R., Chrysanthopoulou, S. A., Lurie, M. N., Nuzzo, J. B., & Papanicolas, I. (2024). Health system resilience during the COVID-19 pandemic: A comparative analysis of disruptions in care from 32 countries. *Health Services Research*, Dec; 59(6), e14382. https://doi.org/10.1111/1475–6773.14382. Epub 2024 Sep 18. PMID: 39295092; PMCID: PMC11622287

Liu, Y., Lee, J. M., & Lee, C. (2020). The challenges and opportunities of a global health crisis: The management and business implications of COVID-19 from an Asian perspective. *Asian Business & Management*, 19(3), 277–297. https://doi.org/10.1057/s41291-020-00119-x. Epub 2020 May 12. PMCID: PMC7216126

Luqmani, Y. A., & El Hashim, A. (2022). The COVID-19 pandemic: A health crisis managed or a panic response with disastrous future consequences? *Medical Principles Practice*, 31(1), 1–10. https://doi.org/10.1159/000520258. Epub 2021 Oct 18. PMID: 34662884; PMCID: PMC8805053

Morcillo Serra, C., Aroca Tanarro, A., Cummings, C. M., Jimenez Fuertes, A., & Tomás Martínez, J. F. (2023). Impact on the reduction of CO_2 emissions due to the use of telemedicine. *Scientific Reports*, 13, Article 14392. https://doi.org/10.1038/s41598-022-16864-2

Mosadeghrad, A. M., Isfahani, P., Eslambolchi, L., Zahmatkesh, M., & Afshari, M. (2023). Strategies to strengthen a climate-resilient health system: A scoping review. *Global Health*, 19, 62. https://doi.org/10.1186/s12992-023-00965-2

OECD (2023). *Ready for the next crisis? Investing in health system resilience, OECD Health Policy Studies*. OECD Publishing. https://doi.org/10.1787/1e53cf80-en

OECD. (2024). *Infrastructure for a climate-resilient future*. OECD Publishing. https://doi.org/10.1787/a74a45b0-en

Olivia, S., Gibson, J., & Nasrudin, R. (2020). Indonesia in the Time of Covid-19. *Bulletin of Indonesian Economic Studies*, 56(2), 143–174. https://doi.org/10.1080/00074918.2020.1798581

Öncü, M. A., Yıldırım, S., Bostancı, S., & Erdoğan, F. (2021). The effect of COVID-19 pandemic on health management and health services: A case of Turkey. *Duzce Medical Journal*, 23(S1), 61–70.

Palmeiro, Y., Reneses, T. P., Velényi, E., Guanais, F., & Cristian Herrera, C. (2023). *Health systems resilient to climate change, an opportunity for sustainable human development*. https://blogs.worldbank.org/en/latinamerica/health-systems-resilient-climate-change

Pohjosenperä, T., Kotavaara, O., & Juga, J. (2022). *CO2 emissions in healthcare logistics: Assessing the effects of green logistics initiatives*. Proceedings of the 8th International Conference on Information Systems, Logistics and Supply Chain (ILS 2022), Lyon, France.

Pollard, C. A., Morran, M. P., & Nestor-Kalinoski, A. L. (2020 Nov 1). The COVID-19 pandemic: A global health crisis. *Physiol Genomics*, 52(11), 549–557. https://doi.org/10.1152/physiolgenomics.00089.2020. Epub 2020 Sep 29. PMID: 32991251; PMCID: PMC7686876

Poynting, M., & Stallard, E. (2024). *How climate change worsens heatwaves, droughts, wildfires and floods*. https://www.bbc.com/news/science-environment-58073295

Rizan, C., Bhutta, M. F., & Reed, M. (2023). The carbon footprint of products used in five common surgical operations: Identifying contributing products and processes. *BMJ Open*, 13(7), e077788. https://doi.org/10.1136/bmjopen-2023-077788

Sagan, A., Webb, E., McKee, M., Greer, S. L., Karanikolos, M., Williams, G. A., Cylus, J., Richardson, E., Waitzberg, R., Lessof, S., & Figueras, J. (2021). Health systems resilience during COVID-19: Lessons for building back better. In A. Sagan, E. Webb, N. Azzopardi-Musco, I. de la Mata, M. McKee, & J. Figueras (Eds.), *World Health Organization*. European Observatory on Health Systems and Policies, Regional Office for Europe. https://iris.who.int/handle/10665/348493

Salam, A., Wireko, A. A., Jiffry, R., Ng, J. C., Patel, H., Zahid, M. J., Mehta, A., Huang, H., Abdul-Rahman, T., & Isik, A. (2023). The impact of natural disasters on healthcare and surgical services in low- and middle-income countries. *Annals of Medicine and Surgery* (Lond). Jul 6; 85(8), 3774–3777. https://doi.org/10.1097/MS9.0000000000001041. PMID: 37554857; PMCID: PMC10406090

Schwerdtle, P. N., Ngo, T. A., Hasch, F., Phan, T. V., Quitmann, C., & Montenegro-Quiñonez, C.A. (2024). Climate change resilient health facilities: A scoping review of case studies in low and middle-income countries. *Environmental Research Letters*, 19, 7.

Sladek, R. M., Gaffney, K., & Braithwaite, J. (2024). Strategies and tactics to reduce the impact of healthcare on climate change: Systematic review. *BMJ*, 387, e081284. https://doi.org/10.1136/bmj-2024-081284

Smith, P. C., Anell, A., Busse, R., Crivelli, L., Healy, J., Lindahl, A. K., Westert, G., & Kene, T. (2012). Leadership and governance in seven developed health systems. *Health Policy*, 106(1), 37–49.

Thomas, S., Sagan, A., Larkin, J., Cylus, J., Figueras, J., Karanikolos, M. (2020). *Strengthening health system resilience: Key concepts and strategies*. European Observatory on Health Systems and Policies. https://apps.who.int/iris/bitstream/handle/10665/332441/Policy-brief%2036-1997-

UN. (2012). *UN System Task Team on the Post-2015 UN Development Agenda, Disaster Risk and Resilience*. https://www.un.org/en/development/desa/policy/untaskteam_undf/thinkpieces/3_disaster_risk_resilience.pdf

UN. (2020). *Policy Brief: COVID-19 and Universal Health Coverage*. https://unsdg.un.org/resources/policy-brief-covid-19-and-universal-health-coverage

UN. (2024). *The Sustainable Development Goals Report*. https://unstats.un.org/sdgs/report/2024/The-Sustainable-Development-Goals-Report-2024.pdf

UNDRR. (2025). *United Nations Office for Disasters Risk Reduction*. Definition: Early Warning System. https://www.undrr.org

USGCRP. (2016). Impacts of climate change on human health in the United States: A scientific assessment. In A. Crimmins, J. Balbus, J. L. Gamble, C. B. Beard, J. E. Bell, D. Dodgen, R. J. Eisen, N. Fann, M. D. Hawkins, S. C. Herring, L. Jantarasami, D. M. Mills, S. Saha, M. C. Sarofim, J. Trtanj, & L. Ziska, (Eds.), *U.S. Global change research program*. (pp. 312). http://dx.doi.org/10.7930/J0R49NQX

WHO & IAHO. (2024). *Climate change is impacting health in Africa: Analytical fact sheet*. https://files.aho.afro.who.int/afahobckpcontainer/production/files/iAHO_Climate_change_in_health_Fact_Sheet-April_2024.pdf

WHO, (n.d.a). *Climate Change and Health*. https://www.who.int/teams/environment-climate-change-and-health/climate-change-and-health/country-support/building-climate-resilient-health-systems

WHO, UNICEF, & World Bank Group. (2022). *State of the world's drinking water: An urgent call to action to accelerate progress on ensuring safe drinking water for all*. World Health Organization. https://www.who.int/publications/i/item/9789240060807

WHO. (2010). *Key components of a well functioning health system*. https://www.paho.org/derechoalaSSR/wp-content/uploads/Documentos/Bloques-Basicos-de-un-Sistema-de-Salud-OMS.pdf

WHO. (2012). *Climate change and health vulnerability and adaptation assessment*. https://www.who.int/publications/i/item/9789240036383

WHO. (2013). *Climate change attributable DALYs* ('000), https://www.who.int/data/gho/data/indicators/indicator-details/GHO/climate-change-attributable-dalys-(000)

WHO. (2015), *Health workforce 2030: Towards a global strategy on human resources for health*.

WHO. (2017). *Emergency response framework* ERF (2nd ed.).

WHO. (2020). *WHO guidance for climate-resilient and environmentally sustainable health care facilities*. World Health Organization. Licence: CC BY-NC-SA 3.0 IGO.

WHO. (2022a). *Alliance for Transformative Action on Climate and Health* (ATACH), https://www.who.int/initiatives/alliance-for-transformative-action-on-climate-and-health/

WHO. (2022b). *Measuring the climate resilience of health systems*. https://www.who.int/publications/i/item/9789240048102

WHO, (2023a). *Advice for the public: Coronavirus disease* (COVID-19), https://www.who.int/emergencies/diseases/novel-coronavirus-2019/advice-for-public

WHO. (2023b). *Universal health coverage* (UHC), https://www.who.int/news-room/fact-sheets/detail/universal-health-coverage-(uhc)

WHO. (2024). *Climate change is impacting health in Africa: Analytical fact sheet.* WHO Regional Office for Africa. https://files.aho.afro.who.int/afahobckpcontainer/production/files/iAHO_Climate_change_in_health_Fact_Sheet-April_2024.pdf

WHO. (n.d.b) *Global Health Workforce statistics database*, https://www.who.int/data/gho/data/themes/topics/health-workforce

WHO. (n.d.c). *World Health Statistics 2024, World health statistics 2024: Tables of health statistics by country and area*, WHO region and globally, https://www.who.int/data/gho/whs-annex/

WHO. (n.d.d). *Emergency cycle.* https://www.who.int/europe/emergencies/emergency-cycle

World Bank Group. (2024). *Health and Climate Change.* https://www.worldbank.org/en/topic/health/brief/health-and-climate-change

World Bank. (2024a). *The Cost of Inaction: Quantifying the Impact of Climate Change on Health in Low- and Middle-Income Countries.*

World Economic Forum. (2024b). *Quantifying the Impact of Climate Change on Human Health.* Insight Report, January 2024.

World Health Organization. (2023b). *Operational framework for building climate resilient and low carbon health systems.* https://www.who.int/publications/i/item/9789240081888

Yıldırım, D. Ç., Yıldırım, S., & Bostancı, S. H. (2024a). Anthropogenic impacts in the Himalayas and the sustainable development goals (SDGs). In A. Borthakur, & P. Singh (Eds.), *The Himalayas in the Anthropocene.* Springer. https://doi.org/10.1007/978-3-031-50101-2_12

Yıldırım, S., & Kantarcı, T. (2022). A Review on sustainability policies of businesses: Recycling and waste reduction. *Journal of Recycling Economy & Sustainability Policy*, 1(1), 1–9. Retrieved from https://respjournal.com/index.php/pub/article/view/1

Yıldırım, S., Sevik, N., Kandpal, V., & Yıldırım, D. C. (2024b). The role of green brands on Achieving 2030 sustainable development goals (2030 SDGs). In K. Kankaew, S. Chaudhary, & S. Widtayakornbundit (Eds.), *Contemporary management and global leadership for sustainability* (pp. 141–162). IGI Global Scientific Publishing. https://doi.org/10.4018/979-8-3693-1273-5.ch009

Yıldırım, S., Yildirim, D. C., & Calıskan, H. (2020). The influence of health on economic growth from the perspective of sustainable development: A case of OECD countries. *World Journal of Entrepreneurship, Management and Sustainable Development*, 16(3), 181–194. https://doi.org/10.1108/WJEMSD-09-2019-0071

Zurl, H., Qian, Z., Stelzl, D. R., Dagnino, F., Korn, S. M., Labban, M., Lipsitz, S. R., Leitsmann, M., Ahyai, S., Ellimoottil, C., Loeb, S., Iyer, H. S., Trinh, Q.-D., & Cole, A. P. (2024). Carbon emissions from patient travel for health care. *JAMA Network Open*, 7(3), e240664. https://doi.org/10.1001/jamanetworkopen.2025.2513

Index